Os ... its

Thro ... ents

070739

TAB Books
Division of McGraw-Hill, Inc.
Blue Ridge Summit, PA 17294-0850

FIRST EDITION
FIRST PRINTING

© 1994 by **TAB Books**.
TAB Books is a division of McGraw-Hill, Inc.

Library of Congress Cataloging-in-Publication Data

Carr, Joseph J.
    Mastering oscillator circuits through projects and experiments /
by Joseph J. Carr.
        p. cm.
    ISBN 0-8306-4067-3   ISBN 0-8306-4066-5 (pbk.)
    1. Oscillators, Electric—Experiments.   2. Electronic circuits-
-Experiments.   I. Title.
TK7872.O7C29 1993
621.3815'33—dc20   92-42360
                                                    CIP

Aquisitions Editor: Roland S. Phelps
Editorial team: Joanne Slike, Executive Editor
            Lori Flaherty, Managing Editor
            Kenneth M. Bourne, Editor
Production Team: Katherine G. Brown, Director
            Patsy Harne, Layout
            Tina M. Sourbier, Typesetting
            Ollie Harmon, Typesetting
            Kelly S. Christman, Proofreading
            Stephanie Myers, Computer Illustrator
            Joann Woy, Indexer
Design Team: Jaclyn J. Boone, Designer
            Brian Allison, Associate Designer   EL1
Cover Design: Holberg Design, York, Pa.   4198

Contents

Preface

This book is another of McGraw-Hill's *Mastering Electronics* volumes. The concept of the series, and the design of this book, is to present electronics technology simply with special emphasis on the "doing" aspects. The learn-by-doing approach is approved by educators and, perhaps more important, by readers of earlier volumes in the series. It's certainly worked for me over the years.

In this book, you will find information on oscillator circuits, multivibrators, and waveform generators. You will be able to design and build both simple and complex versions of these circuits, from frequencies close to dc—or "one-shot" in one case—to the very high frequency (VHF) region of the radio spectrum.

Although emphasis is on performing experiments and building projects as a means of learning the subject matter, tutorial chapters are also included on specific elements needed to understand following chapters. For example, chapters are included on resistance-capacitance (RC) network behavior and integrator and differentiator circuits.

The circuits can be used as stand-alone signal sources, workbench signal generators, radios, instruments, and many other applications. Just how broad the applications spectrum proves to be is up to your own imagination.

1
Doing electronics

THIS BOOK IS ABOUT OSCILLATORS AND OTHER WAVEFORM-GENERATOR CIRCUITS. Such circuits are used to generate signals in the electronics field. But before getting to the meat of the subject, let's look at some background material. Electronics is historically based in the radio industry and the primitive electrical industry that existed at the turn of the century. The first radio receivers were not terribly sensitive. They used simple circuits and unamplified devices such as Branly coherers and galena crystals for detecting electromagnetic signals. A critical need at that time was for a better detector and, above all, an *amplifier* device. The device that provided both of these functions was the triode vacuum tube—the first true electronic component.

Until World War II, the electronics industry was wedded to radio. Many of the components, circuits, and techniques of radio were equally useful for other purposes as well. For example, the amplifier could be used for consumer audio, motion picture, and other entertainment products, and also for electronic instruments and recording devices. The first scientific and medical instruments appeared circa World War I, and by World War II were an established fact. Indeed, they represented a completely new industry separate from radio.

World War II presented the electronics industry with an unprecedented opportunity for advancement. Many new military applications of electronics were invented, and many of these translated directly into civilian technology after the war: communications, radar, control systems, and many others. During the war, the first solid-state signal diodes were created for the VHF through microwave regions of the electromagnetic spectrum. These advances led directly to the invention of the transistor in 1948.

The transistor was invented in the late 1940s by a team of three physicists at Bell Laboratories, a research arm of American Telephone and Telegraph. One form of the device was actually postulated in scientific circles prior to the war by one of the three inventors, but it took the metallurgy research effort of the war years to produce the purity of semiconductor materials needed to produce these new devices.

Perhaps one of the most momentous decisions of the electronics era was the decision by AT&T to license the transistor to other companies. While arguably in their own best interest, for it spurred on the new technology at an unprecedented pace, this decision opened the solid-state era at a much earlier date than would have been otherwise possible. Within 10 to 12 years, vacuum tubes had disappeared from many areas of electronics technology.

The first transistor radios appeared in 1954, with most of the first models being portable AM radios. The low power consumption and low voltage requirements of transistors made them ideal for portable radios. Motorola produced a universal car radio using transistor technology in 1956, but it took until 1962 for the first transistor car radio to be produced for a major automobile manufacturer. That year, the Delco Radio division of General Motors produced all-transistor radios for General Motors. From 1957 to 1962 (or 1963 for other brands), the common design of car radios used low-voltage vacuum tubes for all stages except the AM detector and audio stages, where PN diodes and transistors were used. By 1963, however, all car radios were transistorized, including AM/FM models.

When I first became interested in electronics as a high-school student in the late 1950s, one of my fondly remembered teachers predicted that semiconductor manufacturers were working on technology that would permit putting an entire audio-amplifier chain, or even an entire radio, on a single crystal chip not larger than a thumbnail. Although that visionary prediction seemed a little on the wild side in 1959, by 1969 it was a fact, and devices were in production in home, portable, and car radio receivers.

The prediction mentioned by my first electronics teacher was the invention of the *integrated circuit*, or *chip*. This device integrated onto a single monolithic crystal of semiconductor material all the transistors, diodes, and resistors needed to make a circuit. Some modern circuits also include inductances and capacitances, but only at very high frequencies. These components, built onto a small crystal, are very dense, so the size of electronic devices can be reduced considerably.

Digital electronics also progressed at a tremendous pace. The modern computer would not be practical for most users without the semiconductor industry. Digital circuits obey the rules of the binary (base 2) arithmetic system. In these devices, the numbers 0 and 1 are presented by two different voltage levels, all other voltage levels being prohibited. The first digital devices in large-scale production were the *resistor transistor logic* (RTL) and *diode transistor logic* (DTL). In the 1960s, the now ubiquitous *transistor transistor logic* (TTL) devices were produced. The TTL devices were followed soon thereafter by *complementary metal oxide semiconductor* (CMOS) devices. These latter devices required such low power that many portable devices that were previously impossible became easy.

One of the startling aspects of integrated electronics that has been repeated endless times is the well-known "learning curve" phenomenon. The price of IC devices typically starts out very high when the technology is new and then drops rapidly. For example, the μA-709 operational amplifier was more than $120 in the early 1960s (when dollars were bigger!), but now costs less than 50 cents. The μA-703 RF/IF amplifier cost $15 in the middle to late 1960s when I was servicing electronic equipment, but now costs two-bits, when you can find it. Similarly, I paid $10 for a 741 operational amplifier in the mid-1960s, but now they are 10 for $3 in commercial, plastic 8-pin DIPs. The low cost of these devices makes it reasonable for hobbyists, amateurs, and students to experiment with ICs and use them in actual, practical projects.

Today, electronics includes a bewildering variety of functions, from radio broadcasting, communications of many types, control systems, instrumentation, and many thousands of other applications. Even your automobile is dependent on electronics. While older cars had only a radio as an electronic device, modern cars are run on central computers and have dozens of other electronic devices on board. Even the ignition system is an electronic circuit.

This book is about *oscillators* and, more specifically, *waveform generators*. You will first be introduced to electronic components and electronic construction practices. You will also be introduced to the electronic instruments needed to work the experiments or build the projects. Indeed, I recommend very strongly that newcomers to electronics build or buy a number of electronic instruments for their own use.

This book is intended to be open-ended; for it is hoped you will pursue electronics in ever-widening circles. TAB offers an outstanding array of electronics textbooks, most quite suitable for either self-study or classroom use, from the most elementary levels through the intermediate and most advanced levels. Once you master a level of difficulty on any given topic in electronics, you might want to further explore the same topic, or explore another topic. Practical books can help you in gaining this deeper understanding, so you might want to get a copy of the publisher's catalog.

If I, as the author, ever have the pleasure of visiting your lab or workshop, then I hope to find a copy of this book on the workbench—bent, dirtied, and well-marked. It is no dishonor to the author of a "how-to" book to find that readers have been using the book "where the solder meets the hot tip . . . on the bench."

Workbench skills

Electronics is clearly a very practical field. If the theory is not somehow converted to practice—hardware on the bench—then it is a bit of a waste of time. Much of this book is about electronic theory, although written from a practical perspective. In the rest of this chapter, we'll look at the real down-and-dirty workbench aspects of electronics.

Power supplies for laboratory experiments

You might perform laboratory experiments both independently and as part of a formal class. Guidelines for experiments are given at various points throughout the book. A power supply must be selected (or built) to work these experiments. All power supplies can be worked with batteries, although you might prefer to buy or build your own power supply.

Unless otherwise specified, the experiments in this book are designed to use +12 V (or ± 12 V) dc from a regulated power supply. The power supply selected should offer either a single bipolar power supply or two independent 12-Vdc supplies that are not ground referenced. The "nongrounded" feature allows you to create a bipolar supply by connecting the positive output terminal of one supply to the negative output terminal of the other (see Fig. 1-1).

Desirable features to have on a bench power supply include:

- Output voltages fixed at ±12 Vdc (or ±15 V) or output voltages adjustable from 0 to greater than 12 Vdc.
- Current available from each polarity, not less than 100 mA.
- Metered outputs (I and V).
- Voltage regulation.
- Current-limiting for output short-circuit protection.
- Overvoltage protection.

Although not all good dc power supplies have all these features, those that do are clearly superior for most laboratory applications. All the experiments in this book can be run on any dc power supply that substantially meets these requirements.

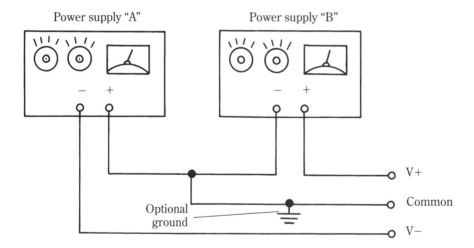

1-1 Use of two power supplies to form bipolar dc supplies for op amps and other ICs.

Construction media

Electronic projects must be constructed on some sort of base, or media. Although novices often build their first few projects "rat's nest style" (i.e., in a three-dimensional "wire sculpture" form), more experienced people use a chassis, printed circuit board, perforated board, or some other media.

Succesfully performing experiments and designing or debugging circuits on a professional level requires certain skills and knowledge of construction practices. In this section, we look at some of these practical matters.

There are various levels of construction practice, of which we can identify four major categories:

1. Partial breadboard
2. Complete breadboard
3. Brassboard
4. Full-scale development model

The two breadboard categories involve using special construction methods to check circuit validity and make certain preliminary measurements. The breadboard is in no way usable in actual equipment, but rather is a bench or laboratory device. The difference between partial and complete breadboards is merely one of scale. The partial breadboard is used to check out a partial circuit, or a small group of circuits. The complete breadboard is a full-up circuit and is more extensive than the partial breadboard. There are fundamental differences in the type of breadboarding hardware used for both types, especially in large circuits.

A common type of breadboard hardware is shown in Fig. 1-2. The device in Fig. 1-2 is an "IC pin socket" breadboard. Some pin-socket breadboards offer two or three built-in dc power supplies with the following typical specifications: +12 V at 100 mA, –12 V at 100 mA, and +5 V at 1.5 A (for TTL digital-logic IC devices).

Located between the large multipin IC sockets are bus sockets, in which all pins in a given row are strapped together. These sockets are generally used for power distribution and common ("ground") buses. Interconnections between sockets, power sources, and components are made using #22 insulated hook-up wire. The bared ends of the wires are pressed into the socket holes.

The other method of breadboarding is wire wrapping (Fig. 1-3). Sockets with special square or rectangular wire-wrap pins are mounted on a piece of "universal" printed circuit board. These boards are perforated to accept wire-wrap IC sockets, and usually have "printed" power distribution and ground buses. Using a special tool,

1-2 Electronic construction breadboard.

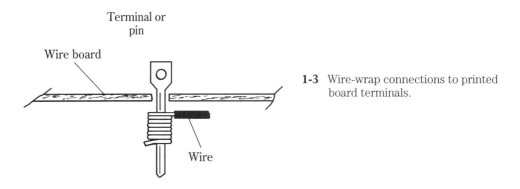

Terminal or
pin

Wire board

1-3 Wire-wrap connections to printed
board terminals.

Wire

the wire interconnects are strapped from point-to-point as needed by the circuit. The wire is wrapped around each contact tightly enough to break through the insulation to make the electrical connection.

Socketed breadboards are typically used for student applications, for partial breadboarding in professional laboratories, and for small design projects that don't justify the cost of a wire-wrap board. Wire wrapping is used for larger, more extensive projects. Although the distinctions among types of breadboards and their respective supporting hardware are quite flexible, they hold true for a wide range of situations.

Brassboards are full-up, model-shop versions of the circuit that plug into the actual equipment cabinet in which the final circuit will reside. Some brassboards are wire wrapped, but most are printed circuits. The important consideration is that the brassboard is near its final configuration and can be tested *in-situ* in the actual equipment. In contrast, the breadboard is purely a laboratory model. Brassboards are usually made using model-shop handwork methods and not routine automated production methods. The brassboard differs from the final version in that it might contain "dead bug" and "kluge card" modifications in which components are informally mounted on the board for testing circuit changes.

The full-scale development (FSD) model is the highest preproduction model. It might look very much like the first-article production model that is eventually made. It is intended for field tests of the final product. For example, an FSD model of a two-way land-mobile radio transceiver might be mounted in an actual vehicle and used for its intended purpose in tests or field trials. For aircraft electronics, the FSD model must be built and tested according to flight-worthiness criteria. The FSD model is built as near to regular production methods as possible.

Breadboarding principles

Several principles should be followed when breadboarding electronic circuits. While it is possible to get away with ignoring these rules, they constitute good practice and ignoring them is risky. The rules are:

- Insert and remove components only with the power turned off.
- Wiring and changes to wiring are done with the power turned off.
- Check all wiring prior to applying power for the first time.

Power distribution and ground wiring is always done first and then checked before any other wiring is done. NOTE: In many schematics, the $V-$ and $V+$ wiring is omitted, but that doesn't mean these connections are not to be made.

Use single-point (AKA *star*) grounding wherever possible. A ground plane or ground bus should not be used unless signal frequencies are low, signal voltage levels are relatively large, and current drain from the dc power supply is low. Otherwise, ground noise and "ground loop" voltage-drop problems develop.

Applying a signal to IC input pins with no dc applied can cause the substrate "diode" to be forward biased, potentially damaging the device. Therefore, don't apply signal until the dc power is turned on; turn off the signal source prior to turning off the dc power supply; and never apply a signal with a positive peak that exceeds $V+$ or a negative peak that exceeds $V-$.

All measurements are to be made with respect to common or ground unless special test equipment is provided. For example, in Fig. 1-4, differential voltage (V_d) is composed of two ground-referenced voltages, V_1 and V_2. Measure these voltages separately and then take the difference between them (i.e., $V_2 - V_1$).

Although it is relatively easy today to make a small printed circuit board using "home brew" techniques, most small projects are constructed (at least initially) on perforated wiring board (usually just called *perf board*). Figure 1-5 shows both RF (Fig. 1-5A) and subaudio (Fig. 1-5B) projects built on perf board. RF projects should be housed in metal shielded chassis, such as those shown in Fig. 1-6.

Designing electronic projects

The process of designing electronic circuits or projects is not an arcane art open only to a few. Rather, it is a logical step-by-step process that can be learned. Like any skill, design skill is improved with practice. So, be forewarned of both excessive expectations "first time at bat" and discouragement if the process did not not exactly work out as planned the first time.

Adopting a design procedure

The procedure you adopt to design might well be different from what is presented here, and that's alright. The purpose of offering a procedure is to *systematize* the

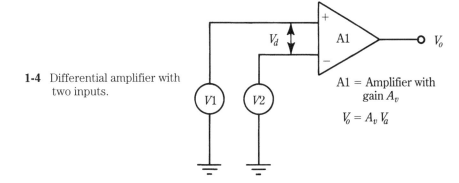

1-4 Differential amplifier with two inputs.

A1 = Amplifier with gain A_v

$V_o = A_v V_d$

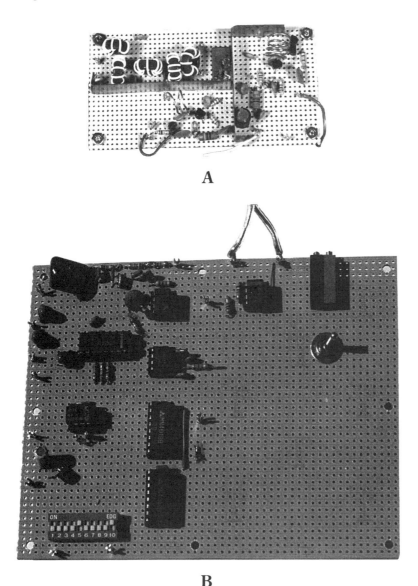

1-5 A. RF project on perfboard. B. Digital project on perfboard.

process. While it is conceivable that an instrument can be designed by a process similar to Brownian motion (as, theoretically, a Tom Clancy novel can be written by an infinite collection of monkeys pounding randomly on an infinite number of PC keyboards), it is the systematic approach that most often yields success.

Designing a product for production and sale follows a similar procedure but involves marketing and production problems as well. The steps in the procedure, some of which are iterative with respect to each other, are:

1. Define and tentatively solve the problem.
2. Determine the critical attributes of the final product. Incorporate the attributes into a specification and a test plan that determines objective criteria for acceptance or rejection.
3. Determine the critical parameters and requirements.
4. Attempt a "block diagram" solution.
5. Apportion requirements (e.g., gain, frequency response, etc.) to the various blocks.
6. Perform analysis and do simulations on the block diagram to test validity of the approach.
7. Design specific circuits to fill in the blocks.
8. Build and test the circuits.
9. Combine the circuits with each other on a breadboard.
10. Test the breadboarded circuit according to a fixed test plan.
11. Build a brassboard that incorporates all changes made in the previous steps.
12. Test the brassboard and correct problems.
13. Design and construct final configuration.
14. Test final configuration.
15. Ship the product.

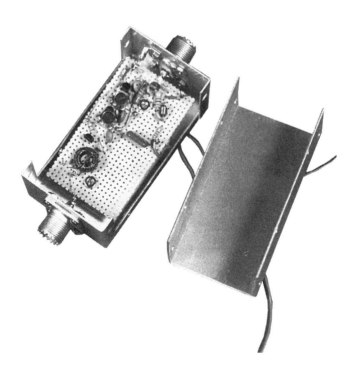

1-6 Shielded aluminum box used for electronic construction.

Solving the (right) problem The purpose of the designer is to solve any problems using analog circuits, digital circuits, a computer, or whatever else is available in the armamentarium. Novices often encounter two related problems.

First, the designer often has a tentative favorite approach in mind before the problem is properly understood. Decisions are made based on what the designer is most comfortable doing. For example, many younger designers are likely to select the digital solution in a knee-jerk manner that excludes any consideration of the analog solution. Both should be evaluated, and the one that best fits the need selected.

Second, be sure you understand the problem being solved. While this advice seems trivial, it is also true that failure to understand the problem often sinks designs before they have a chance to manifest themselves. There are several facets to this problem. For example, engineers naturally tend to think that an elegant solution is complex and large scale. If this mistake is made, the product will likely be overdesigned with too many whistles and bells. It was, after all, designed to solve a much harder problem than was actually presented.

Another aspect to understanding the problem is understanding the end user's *use* for the product. It is all too easy to get caught up in the specification or your own ideas about the job and overlook what the user needs.

For example, a physiologist requested a pressure amplifier that would measure blood pressures over a range of 0 to 300 mmHg (torr). What the salesperson never told the plant was:

1. It would be used on humans (safety and regulatory issues);
2. Blood would come in contact with the diaphragm (cleaning and/or liquid isolation issues); and
3. It would occasionally be used for measuring 1 to 5 mmHg central venous pressures (which implies low-end linearity issues).

Part of the problem in determining the level of complexity or the specific design's function is miscommunication between the end user and the designer. Although miscommunications occur frequently between in-house designers and their "clients," they are probably most common between distant customers and engineers in the plant. Of course, marketing people might never let the engineer and customer get together (either from ignorance or a fear that the salesperson's little lies will surface: "The reason I hate engineers is that, under duress, they tend to blurt out the truth").

The proper role of the designer is to scope out the problem at hand and understand what the circuit or instrument is supposed to do, how and where the user is going to use it, and exactly what the user wants and expects from the product.

Determining crucial attributes This step is basically the fruit of understanding and solving the correct problem. From the solution of the problem one can determine, and write down, a set of attributes, characteristics, parameters, and other indices of the product's final nature.

At this point, a specification that documents what the final product is supposed to do must be written. The specification should be clearly written so others can understand it. A concept or idea does not really exist except in the mind of the originator. According to W. Edwards Deming,* *operational definitions* for the attributes of

*Out of the crisis, W. Edwards Deming, MIT Press.

the product must be created. It is not enough simply to say, "It must measure pressure to a linearity of 1-percent over a range of 0 to 100 psi." Rather, specify a test method under which this requirement can be met. After all, there might be more than one standard for pressure and measurement, and there is certainly more than one definition of linearity. The operational definition serves the powerful function of providing everyone with the same set of rules, leveling the playing field.

Part of this step, and of making an operational definition, is to write a test plan for the final product. It is here that it is determined (and often contractually agreed upon) exactly what the final product will do and the objective criteria of goodness or badness that will be used to judge the product is defined.

Determining crucial parameters and requirements Once the product is properly scoped out, the critical technical parameters needed to meet the test requirements must be determined (and hopefully the user's needs if the test requirements are properly written). Parameters such as frequency response and gain tend to vary in multimode instruments, so one must determine the worst case for each specification item and design for it.

Attempting a block-diagram solution A block diagram is a signals-flow or function diagram that represents stages, or collections of stages, in the final instrument. In large instruments, there might be several indexed levels of a block diagram, each one finer in detail.

Apportioning requirements to the blocks Once the block diagram solution is on paper, system requirements are tentatively apportioned to each block. Gain, frequency response, and other attributes are distributed to each block. Keep in mind that factors such as gain distribution can have a profound effect on dynamic range. Also, the noise factor and drift of any one amplifier can have a tremendous effect on the final performance. It's these types of parameters where critical placement of one high-quality stage may be sufficient that added cost and complexity often arise.

Analyzing and simulating Once the block diagram is set, and the requirements apportioned to the various stages, it is time to analyze the circuit and run simulations to see if it will actually work. A little "desk checking" goes a long way towards eliminating problems before the first prototype. Plug in typical input values and evaluate results on a stage-by-stage basis. Check for the reasonableness of outputs at each stage. For example, if the input signal should drive an output signal to, say, 17 V, and the operational amplifiers are only operated from 5-V power supplies, then something is wrong and must be corrected.

Designing specific circuits for each block At this point, the remainder of the book is most useful to you, because it covers the specific circuits. In this step, fill in the blocks with real circuit diagrams.

Building and testing the circuits At this point, the individual circuits must actually be constructed and tested to ensure they work as advertised (unless, of course, the circuit is so familiar that no testing is needed). Keep in mind that some of your best ideas for simplified circuits might not actually work—and this is the place to find out. Use a benchtop breadboard that allows circuit construction using plug-in stripped-end wires.

Combining the circuits in a formal breadboard Once the validity of the individual circuits is determined, combine them in a formal breadboard. Whether built on a benchtop breadboard or on a prototyping board, make sure the layout is similar to that expected in the final form.

Testing the breadboard The overall circuit is tested according to a formally established objective criteria—the test plan developed earlier in the second step. As problems arise and are solved, make changes and/or corrections and document the results. The main failing of inexperienced designers is improper documentation, even in an engineer's or scientist's notebook.

Building and testing a brassboard version The brassboard is made as close as possible to the final configuration. While breadboarding techniques can be a little sloppy, a brassboard should be a proper printed circuit board. The test criteria should be the same as before, updated only for changes. If problems turn up, they should be corrected prior to proceeding. Keep in mind that the most common problems that occur in leaping from breadboard to brassboard are layout (e.g., coupling between stages), power distribution, and ground noise, or the areas of difference between the two configurations.

Designing, building, and testing the final version Once all problems are known and solved, and changes incorporated, it is time to build the final product as it will be given to the end user. It is at this point that the reputation of the designer is made or broken, because it is here that the product is finally evaluated by the client.

Electronic instruments and their use

Some particularly skilled electronics enthusiasts seem to be able to divine the problem in a circuit, almost if by magic. While it seems like magic to the noninitiate, it is really only a reflection of a large amount of experience remembered only from past jobs. But even the most skilled, seemingly magical electronics technologist will agree that a good selection of test equipment is the key to wringing out circuit problems and repairing faults.

Selecting test equipment depends on the job you want to do. It is assumed you are interested in solid-state circuits, integrated circuits, and dc and relatively low frequency (below 1 MHz) ac circuits. Therefore, the test equipment discussed in this chapter is aimed at the equipment needed to work on these.

For integrated-circuit work, you will need several different types of instruments: multimeter, signal generators, oscilloscope, power supplies, and certain accessories. Let's take a look at these and some practical examples.

Multimeters

A multimeter is an instrument that combines into one package a multirange voltmeter, milliammeter (or ammeter), ohmmeter, and possibly some other functions. A multimeter measures all the basic electrical parameters needed to troubleshoot a receiver.

The earliest multimeters were instruments called *volt-ohm-milliammeters* (VOMs). These instruments were passive devices, and the only power used in them was a battery for the ohmmeter function. Because they are passive instruments, they are still preferred for troubleshooting medium- to high-power radio transmitters.

They are included here, even though not recommended, because they are low cost and available at electronics outfitters such as Radio Shack.

The sensitivity of the VOM was a function of the full-scale deflection of the meter movement used in the instrument. Sensitivity was rated in terms of ohms per volt full-scale. Three common standard sensitivities were: 1000 Ω/V, 20,000 Ω/V, and 100,000 Ω/V. These represented full-scale meter sensitivities of 1 mA, 100 μA and 10 μA.

Voltage was read on the VOM by virtue of a multiplier resistor in series with the current meter, which forms the basic instrument movement. The front-panel switch selected which multiplier resistor, hence which voltage scale, was used.

The earliest form of an active multimeter was the vacuum-tube voltmeter (VTVM), which used a differential balanced amplifier circuit based on dual-triode tubes (e.g., 12AX7) to provide amplification. As a result, the VTVM was more sensitive than the VOM. It typically had an input impedance of 1 MΩ, with an additional 10 MΩ in the probe for a total input impedance of 11 MΩ. Compare this with the VOM that had a different impedance for each voltage scale.

The next improvement in meters was the solid-state meter. Solid-state meters used a field-effect transistor in the front end and other transistors in the rest of the circuit. As a result, these instruments were called *field-effect-transistor voltmeters* (FETVM). The FETVM has a very high input impedance, typically 10 MΩ or more.

Finally, the most modern form of instruments are digital multimeters (DMM) in both hand-held and bench (Fig. 1-7A, B) designs. These instruments are the meter of choice today if you are going to purchase a new model. In fact, it might prove a bit difficult to find a nondigital model anyplace except Radio Shack. The DMM is preferred because it provides the same high input impedance as the FETVM but, because the display is numeric instead of analog, the readout is less ambiguous.

One bit of ambiguity, however, is last-digit bobble. The digital meter only allows certain discrete states, so if a minimum value is between two of the allowed states, then the reading can switch back and forth between the two, especially if noise is present. For example, a voltage might be 1.56456 V, but the instrument is only capable of reading to three decimal places. Thus, the actual reading might bobble back and forth between 1.564 and 1.565 V.

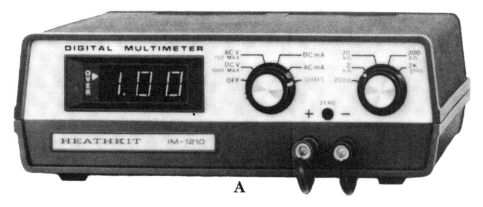

1-7A Digital multimeter. Bench-type model.

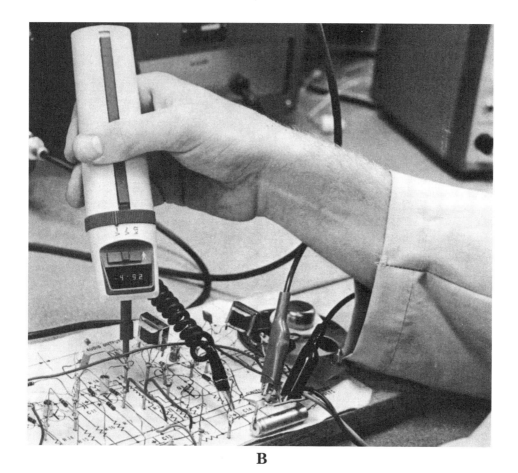

B

1-7B Digital multimeter. Hand-held model.

Last-digit bobble is not really a problem, however, unless you bought an instrument that is not good enough. The number of digits on the meter should be one more than that required for the actual service work. Digits are specified in DMMs in terms of the number of places the meter can read. For example, some instruments are called a 2½ digit model. The ½ digit refers to the most significant digit (on the left side), which can only be 1 or 0 (in which case it is blanked off). Thus, the full scale of this meter is always 199. The position of the decimal point is a function of the range setting.

For many users, the DMM selected can be a 2½ digit model, but if you anticipate a lot of detailed solid-state work, then opt instead for a 3½ digit model. These instruments will read to 1999 on each scale. Oddly, the cost difference between the two is small. A 4½ digit meter is even better but is not strictly necessary for most of you.

Interestingly, DMMs use low voltage for the ohmmeter for compatibility with solid-state equipment. The ohmmeter on other types of instruments will forward bias PN junctions of solid-state devices, and so cannot be used for accurate measurements. However, this same feature of the DMM also means that the normal ohmmeter cannot be used to make quick tests of PN-junction devices such as transistors and diodes. However, some DMMs have a special switch setting that allows this. In some DMMs it is a high-power setting, while in others it is designated by the normal arrow diode symbol.

Another feature of the modern DMM is an aural continuity tester that sounds a *beep* anytime the resistance between the probes is low, i.e., indicating a short circuit exists. This feature is especially useful for testing multiconductor cables for continuity when the actual resistance is not a factor.

Older instruments

As is also true with other equipment, multimeters show up on the used market (especially hamfests) quite often. It is easy to obtain workable, if older, instruments at reasonable prices, especially now that everyone seems to be dumping old analog instruments in favor of the new digital varieties. However, there are some pitfalls.

First, make sure the instrument isn't so old that it uses a 22½ V (or higher) battery in the ohmmeter section. Those instruments were made prior to about 1956 and will blow most transistors or integrated circuits you measure with them. Look out especially for the oversized RCA and Hickock instruments that were once the mainstay of radio service shops.

Second, make sure the instrument does not have a problem that cannot easily be repaired. VTVMs tend to be in good shape, or at least repairable condition. Unfortunately, many VOMs are in terrible shape. If the operator left the meter on the current setting and then measured a voltage . . . pooff! The instrument burns up. When examining an instrument, try to make a measurement on each scale. Alternatively, remove the case and look for charred resistors and burned switch contacts.

In any event, when you obtain a used meter, be sure to take it out of the case and clean the switch contacts and generally spruce it up. Commonly available contact and tuner cleaner will work wonders on a used multimeter as well. A simple cleaning of the switch(s) could remove intermittent or erratic operation.

Signal generators

The purpose of a signal generator is to produce a signal that can be used to troubleshoot, align, adjust, or simply prove the performance of a piece of electronic equipment. Fortunately, only a few basic forms of signal generator are needed. Signal generators for this market come in several varieties, including: audio generators, function generators, and RF signal generators.

Audio generators produce sine waves on frequencies within the range of human hearing (20 Hz to 20 kHz). Some models produce frequencies over a greater range, while others produce square waves as well as sine waves. The standard audio signal generator has a variable amplitude, or level control, and a fixed output impedance of 600 Ω. Audio generators may or may not have an output level meter.

Function generators are much like audio generators but output a triangle waveform in addition to the sine-wave and square-wave signals. Some function generators also produce pulses, sawtooths, and other waveforms as well. Typical function generators operate from less than 1 Hz to more than 100 kHz. Many models have a maximum output frequency of 500 kHz, 1 MHz, 2 MHz, 5 MHz, and, in at least one case, 11 MHz.

Like the audio generator, the standard output impedance is 600 Ω. However, some instruments also offer 50-Ω outputs (standard for RF circuits) and a TTL (digital) compatible output. The latter is a digital output compatible with the ubiquitous transistor transistor logic (TTL or T²L) family of digital logic devices (the ones with either 74xx or 54xx type numbers).

Sweep function generators allow the output frequency to sweep back and forth across a range around the set center frequency. As a result, frequency-response evaluations can be done using the sweep function generator and an oscilloscope.

RF signal generators output signals generally in the range above 20 kHz and typically have an output impedance of 50 Ω. This value of impedance is common for RF circuits except in the television industry, where 75 Ω is the standard. RF signal generators come in various types but can easily be grouped into two general categories: service grade and laboratory grade. For service work, either type is usable.

My own bench has several different signal-generator instruments: the Heath IG-18, a sweep function generator; a Model 80 (2–400 MHz) RF signal generator; a Japanese signal generator that is sold under different brand names; and an elderly Precision Model E-200C that I refurbished after buying it at a hamfest in nonworking condition.

Oscilloscopes

Probably no other instrument is as useful as the oscilloscope, formally called the *cathode ray oscillograph* or CRO, in working on IC circuits. It displays the input signal on a viewing screen provided by a cathode ray tube (CRT). You can look at signals and waveforms instead of just averaged dc voltages, which are measured on the DMM. Figure 1-8 shows a simple service-grade instrument that can be used for most service jobs on AM, FM, and shortwave receivers, as well as most CB and amateur radio transmitters.

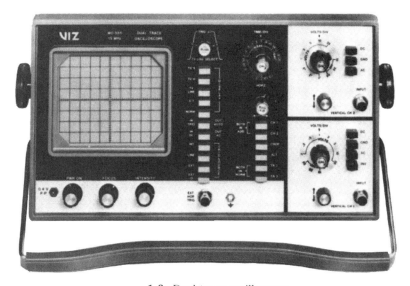

1-8 Dual-trace oscilloscope.

The light on the CRT viewing screen is produced by deflecting an electron beam vertically (Y-direction) and horizontally (X-direction). When two signals are viewed with respect to each other on an X–Y oscilloscope, the result is a Lissajous figure.

The quality of oscilloscopes is often measured in terms of the vertical bandwidth of the instrument. This specification refers to the –3 dB frequency of the vertical amplifier, or amplifiers if it is a dual-beam scope. The higher the bandwidth, the higher the frequency that can be displayed and the sharper the rise-time pulse that will be faithfully reproduced. For ordinary service work, a 5-MHz scope will suffice for most applications. However, there will be times when you need a high-frequency scope, so buy as much bandwidth as you can afford. Although once in the price stratosphere, even 50-MHz models can be bought relatively cheaply—brand new.

dc power supplies

Although often overlooked as "test equipment," the simple dc power supply is definitely part of the test bench. In addition to powering units being repaired that either lack a dc supply or have a defective dc supply, the bench power supply can be used for a variety of biasing and other troubleshooting functions. In chapter 5 you'll learn the basic theory of dc power supplies, and in chapter 3, you'll find construction projects that allow you to work the projects in this book.

Signal tracers

Signal tracers are high-gain audio amplifiers that can be used to examine the signal at various points along the chain of stages in a radio or audio amplifier. A *demodulator probe* permits the signal tracer to "hear" RF and IF signals. The signal tracer is a back-up replacement for the oscilloscope if it is absolutely impossible to obtain a scope. However, the signal tracer is such a poor backup to the scope as to be deemed insufficient. Nevertheless, there are advantages to these instruments and their cost is so low as to be a worthwhile addition to the bench.

2
Introduction to waveforms

THIS BOOK COVERS ELECTRONIC CIRCUITS THAT GENERATE *WAVEFORMS*. VARIOUS circuits create waveforms, typically called *oscillators* or *multivibrators* (either *astable* or *monostable*). This brief chapter discusses the term *waveform*, to prepare for later discussions.

A waveform is a voltage or current function of time. That is, if you graph the voltage between two points or graph the current flowing past a point over specified time, then you basically show a waveform. While this definition lacks a certain mathematical rigor that some could (quite rightly) fault, in the practical context it serves remarkably well.

The mathematically oriented reader (whom we will satisfy in a later chapter) would point out that all continuous waveforms can be described in terms of a fundamental sine-wave frequency (f) in combination with a variety of sine and cosine harmonics of the fundamental frequency ($2f$, $3f$, . . .). That fact is not only mathematically interesting, but also serves to guide us towards methods for shaping waves. You'll learn, for example, that the square wave, which is easily generated, contains a large collection of harmonics of a certain series and phase relationship. By using frequency-sensitive RC, RL, LC, and RLC networks, we can alter the shape of the square wave to a different shape. Musicians know the harmonics as "overtones," and will tell you that different instruments sound different when playing the same fundamental note (e.g., 440 Hz) because of the specifics of the overtones produced by that instrument. Thus, a violin, a piano, and a trumpet can all produce 440 vibrations per second in the air, but sound different because of the harmonics present and their respective phase relationships to the fundamental. Early in the electronic music era before computers were used to make the sounds, complex analog filtering was performed on square, sawtooth, and triangle waveforms to create the various sounds generated by a synthesizer.

Sinusoidal waveforms

The fundamental building block of complex waveforms is the sine wave (Fig. 2-1). This is the waveform of the alternating current (ac) power provided by the power company. When undistorted, the waveform is a "pure" signal because it contains just one frequency. That *frequency* is defined as the reciprocal of the *period* (*T*) of the waveform. The period is the time required for the waveform to make one complete positive excursion followed immediately by one complete negative excursion. The period and frequency are related by:

$$f = \frac{1}{T} \tag{2-1}$$

Where: *f* is the frequency in Hertz (Hz) (once known more descriptively as *cycles per second*)
T is the period in seconds (s)

Example 2-1

Find the frequency of a sine-wave signal that has a period of one millisecond (1 ms). NOTE: 1 ms = 0.001 s.

Solution:

$$f = \frac{1}{0.001 \text{ s}} = 1000 \text{ Hz}$$

Similarly, if you know the frequency and need the period, then switch the *f* and *T* terms in Eq. 2-1 and inverse the calculation.

The time relationships along the baseline of the sine wave can be defined either in terms of actual time (e.g., milliseconds or microseconds), or in terms of *phase angle*. The sine wave has a total of 360° and by definition starts at 0° and finishes at 360°. Because 0° and 360° are the same point on a circle, these points on a sine wave have exactly the same voltage value (zero) and the voltage goes in exactly the same direction (positive). The phase angle at any point might be given in degrees or it might be given in *radians* (there are 2π radians per cycle). Figure 2-1 shows both notations.

Certain relationships are known about the sine wave. When you determine the voltage of a particular sine wave, for example, you could mean any of four different things: *peak voltage*, *peak-to-peak voltage*, *root-mean-squared voltage* (AKA *rms voltage*), or the *instantaneous voltage*. In general, the peak voltage is represented by *V*, and the instantaneous voltage by *v*. Similarly, the peak current is *I*, while the instantaneous current is *i*. To prevent confusion with the other voltages, however, some people designate V_p for the peak voltage, $V_{p\text{-}p}$ for the peak-to-peak voltage, and V_{rms} for the rms voltage (the instantaneous voltage notation remains the lowercase *v*). Because these different values are important for various purposes, it is best to distinguish them from each other.

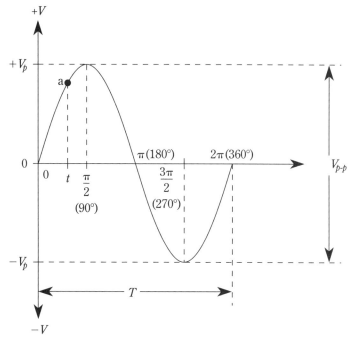

2-1 The sine wave and its key attributes.

What do these different voltages mean? Refer to Fig. 2-1 again. The peak voltage is the voltage difference between the zero volts baseline and either negative or positive maximum (e.g., peak) values. In a perfect sine wave these values are equal magnitude, but opposite polarity from each other. The peak-to-peak voltage is the voltage difference between the maximum negative peak and the maximum positive peak, or:

$$V_{p\text{-}p} = V_{p+} - V_{p-} \qquad\qquad \textbf{(2-2)}$$

When the sine wave is pure, then the peak-to-peak voltage will be twice either peak voltage:

$$V_{p\text{-}p} = 2|V_p| \qquad\qquad \textbf{(2-3)}$$

The root-mean-square (rms) voltage requires a brief explanation. Some people define the rms voltage as the time-average voltage, but that is not quite correct. After all, the time-average voltage of a perfect sine wave over one complete cycle is zero because the positive and negative halves of the sine wave cancel each other. Even when only one half-cycle is considered, the rms value is still not the true average. For a sine wave (only), the average voltage over a half cycle is 0.636 V_p, while the rms voltage is 0.707 V_p. The rms value expresses something a little bit different.

The rms voltage is best described in terms of an analogy that relates the amount of heat dissipated by an equivalent dc voltage in a resistance. If you measure the amount of heat dissipated by the alternating current in a standard resistance when a certain peak voltage is present, and then compare it to the dc voltage that produces the same degree of heating in the same resistor, the dc voltage is the same value as the rms voltage of the ac waveform. For sine waves only, the complex calculus that rigorously defines rms voltage can be reduced to:

$$V_{rms} = \frac{V_p}{\sqrt{2}} = 0.707\ V_p \tag{2-4}$$

Example 2-2

What are the peak-to-peak and rms values of a perfectly symmetrical ac sine wave with a peak voltage (V_p) of 12 volts?

Solution:

$$\begin{aligned} V_{p\text{-}p} &= 2V_p \\ &= (2)\ (12\ \text{V}) \\ &= 24\ \text{V} \\ V_{rms} &= 0.707\ V_p \\ &= (0.707)\ (12\ \text{V}) \\ &= 8.484\ \text{V} \end{aligned}$$

The instantaneous voltage (v) is the voltage at any specified point (e.g., point "A" in Fig. 2-1), and can be found by:

$$v = V_p \sin(\theta) = V_p \sin(2\pi\omega t) \tag{2-5}$$

The term *cosine wave* is sometimes seen in circuit descriptions, but is somewhat misleading. The term has no meaning except in relationship with a sine wave. Specifically, a cosine wave is sinusoidal, like the sine wave, but is displaced from a reference sine wave by 90 degrees (or $\pi/2$ radians). Remove the reference sine wave or choose a phase angle other than 90 degrees, and the cosine wave becomes meaningless. Where there is no sine wave, the cosine wave becomes the sine wave by default, and where the angle is other than 90 degrees, we simply have two sine waves of the same frequency and some random phase relationship. Regardless, however, the so-called cosine wave obeys all the rules for sine waves, so is considered identical unless referenced to some specific sine wave (Fig. 2-2).

More could be said about the sine wave, for it has a similar natural elegance as the normal distribution curve. However, this discussion is sufficient for our present context. Next, we'll turn our attention to other waveforms that are not sinusoidal, although all can be generated by summing up sine and cosine harmonics and a sine fundamental.

Square waves

The square wave has two quasistable states; i.e., it snaps to a particular voltage and remains there for awhile, and then snaps back to the other of two available voltages

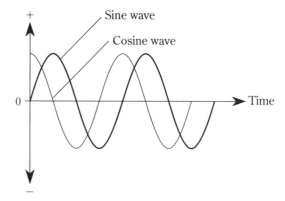

2-2 Sine wave and cosine wave.

and remains there for awhile (see Fig. 2-3A). A perfectly symmetrical square wave, as shown in Fig. 2-3A, has a maximum positive voltage that is equal in magnitude to the maximum negative voltage. Similarly, the durations of the positive and negative segments are equal.

Some square waves are not symmetrical about the base line, so the maximum positive and negative voltages are not equal. These square waves are said to have a *dc component* equal to the difference between $+V_{max}$ and $-V_{max}$. The dc component is also sometimes referred to as the *dc offset* of the waveform.

A unipolar square wave is shown in Fig. 2-3B. Although the purist can argue that only symmetrical, bipolar square waves truly qualify for the designation, common practice also allows either positive or negative unipolar square waves that designation. The positive unipolar variety shown in Fig. 2-3B has a base line at zero volts and a peak value of $+V_{max}$. A negative unipolar square wave would be quite similar, except that it would have a negative peak value. When the two unipolar levels are 0 V and +5 V (actually $V_{max} > +2.4$ V), the wave is a *TTL compatible* square wave; that is, it can be used with transistor transistor logic (TTL) digital integrated circuits.

Pulses

A pulse is a single waveform, usually square, and possibly may be repeated (as in Fig. 2-4). When a circuit generates a single square pulse in response to a trigger signal, it is said to be a *one-shot* circuit (or, more formally, a *monostable multivibrator*). When the waveform is repeated, it forms a *pulse train*. The difference between the square wave and a pulse train is practically a matter of semantics. The pulse train generally has on times (T_{on}) that are short compared with the off times (T_{off}), whereas in a true square wave $T_{on} = T_{off}$. For practical purposes, however, there is no real break point when the times are close but not equal. And since the circuit behavior is similar for both forms of wave, the point is essentially moot at the level where we are working.

Triangle waves

A triangle waveform (Fig. 2-5) can be bipolar (as shown) or unipolar. In either case, the waveform starts at some negative peak (or zero if unipolar), rises linearly to a pos-

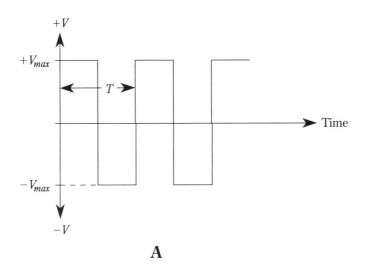

A

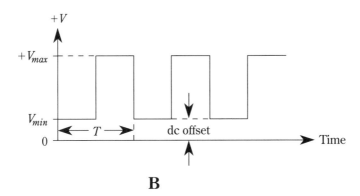

B

2-3 A. Normal ideal square wave. B. Square wave with dc offset.

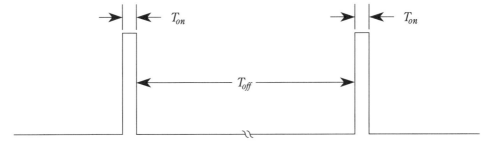

2-4 Pulse waveform.

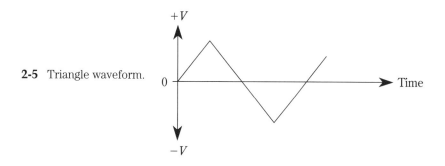

2-5 Triangle waveform.

itive peak, and then reverses direction and drops linearly back to the original value. A triangle wave can be fully bipolar, unipolar negative, unipolar positive, or bipolar asymmetrical (i.e., $-V_{max}$ {does not equal} V_{max}). A triangle wave can be made by low-pass filtering, also called *integrating* a square wave of the same frequency.

Sawtooth waves

A sawtooth waveform (Fig. 2-6) is like half a triangle wave: it starts at some voltage and rises linearly to a peak. When it reverses direction, it drops suddenly to the original value. Sawtooth waveforms can exist in the same combinations of bipolar and unipolar as the other waveforms.

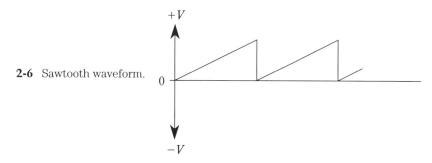

2-6 Sawtooth waveform.

Conclusion

Now that we've learned some facts about waveforms, let's take a look at some oscillator, mulivibrator, and other waveform-generator circuits.

3
dc power sources

ELECTRONIC EXPERIMENTERS FREQUENTLY NEED A SOURCE OF LOW-VOLTAGE direct-current (dc) electrical power for their projects, especially if solid-state electronic devices are being used. You might build simple, but effective electronic circuits based on easy-to-apply devices such as operational amplifiers or 555 integrated-circuit timer chips. Such circuits typically require either a monopolar dc source in the +4.5- to +15-V range, or a bipolar supply in the ±4.5- to ±15-V range.

This chapter includes a construction project for a versatile workbench and laboratory dc power supply that is based, in part, on a commercially available kit. Unless otherwise noted, the experiments and projects in this book were tested on this bench power supply. It is *bipolar*, meaning that it has two separate outputs: one provides a positive output voltage and the other provides a negative output voltage. Furthermore, the two output voltages can independently be either fixed or adjustable over the range ±1.25 to ±17 V. Up to 500 mA (0.5 A) of direct current can be drawn from the power supply. The power supply is *voltage regulated*, so it can be used in a variety of critical electronic or instrumentation applications.

You might want to wire the circuit on perforated circuit board such as Vectorboard® (available at electronic-supplies outlets), but you are well advised to buy the kit. This project is based on the Jameco Electronics* Model JE-215 Dual Power Supply Kit. This kit contains a printed circuit board, and all the small parts that are mounted on the board. (Additional parts are needed for the cabinet, but more of that later.) In addition, the kit includes detailed assembly instructions for the novice electronic builder.

Experienced builders may also want to buy the Jameco JE-215 kit. First, the kit is less difficult to assemble than point-to-point "rat's nest" wiring on perforated board. Second, when I priced the individual components in a catalog, even at dis-

---

*1355 Shorway Road, Belmont, CA 94002. (415) 592-8097.

count prices, the total cost was from 0.9 to over two times the retail price of the kit, depending on the sources of the parts. In other words, there is no real savings under any circumstance where the parts must be bought, and it may well be twice as expensive to supply your own parts—and you don't have the advantage of a professional printed circuit board. The parts list is shown at the end of this chapter.

The power-supply circuit

The circuitry for the dc power supply (Fig. 3-1) is simple and straightforward. A step-down transformer (T1) reduces the 115-V alternating current received from the wall outlet in your home to the low voltage level required for the dc power supply. A pair of *rectifier diodes* (CR1 and CR2) converts the alternating current to an impure form of dc (called *pulsating dc*), while *filter capacitors* (C1 and C2) make the rectifier output nearer to pure dc. The working of these components is beyond the scope of this book, but many basic electronics books cover these topics in depth.

A pair of integrated-circuit *voltage regulators* (IC1 and IC2) stabilizes the output voltage levels under a variety of different conditions. The LM-317T (IC1) is the positive voltage regulator, while the LM-337T (IC2) is the negative voltage regulator. A typical package for these devices is shown in Fig. 3-2. It is a plastic "power transistor-like" package with a metal mounting tab and three electrical connection pins (labeled 1, 2, 3). Note in Fig. 3-1, however, that the pins have different functions on the two different types of regulators.

The LM-317T and LM-337T devices are called *adjustable three-terminal integrated-circuit voltage regulators*. The adjustable feature is programmable by the ratio of two resistors: R2/R3 on IC1, and R4/R5 on IC2. In the Jameco kit, the variable resistors (R3 and R5) set the required output voltage. Jameco supplies these parts as small "trimmer" potentiometers (variable resistors) that mount on the printed circuit board. You might prefer to replace these parts with separate output controls that are mounted on the front panel.

The output terminals for the dc-power-supply printed circuit board (see Fig. 3-3) are a set of four #6 machine screws and nuts mounted on one end of the board. The two center terminals are connected together to form a common electrical connection that serves as the ground points for the two supplies. The other two screws are the negative (−) and positive (+) output voltage terminals.

Follow the printed Jameco instructions when assembling the printed circuit board, except for one point. The integrated-circuit voltage regulators (IC1 and IC2) should be mounted slightly different than shown in the instructions. Figure 3-4 shows the method recommended in the instructions: the two outer pins (1 and 3) are bent away from the center line. This method has the potential for causing internal damage to the voltage regulator because of the stress placed on the internal connections. I prefer the method shown in Fig. 3-4B. Be careful to leave enough length to pass through the printed circuit board. Nonetheless, the method shown in Fig. 3-4A results in a healthier future for the regulator.

At the ac-input side of the power-supply circuit four external components have been added (i.e., not included in the kit). The ON/OFF power switch (S1A and S1B)

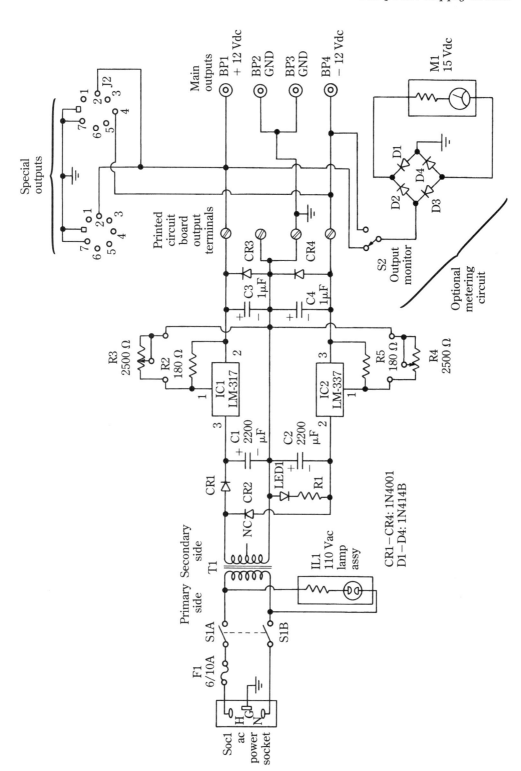

3-1 dc power supply circuit.

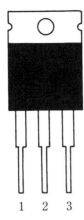

3-2 Plastic-package voltage regulator.

1   2   3

is actually a dual switch so that both sides of the ac power line are opened up when the circuit is turned off. A single switch will work, but there are safety reasons for using two switch sections operated by a single handle. Buy either a double-pole single-throw (DPST) or double-pole double-throw (DPDT) switch that is rated for 125 Vac, or higher.

The ac power line from the wall socket has three wires (don't use a two-wire ac power cord—those are not always safe in electronic instruments!) labeled *hot*, *neutral*, and *ground*. The hot and neutral are connected through the fuse and switch to the primary side of the power transformer (T1). The fuse protects the circuit in case of a catastrophic failure. Perhaps more important, however, is the protection to your home in case of such a failure. Not using a fuse can cause a fire if the power supply shorts and the power cord heats up enough from overcurrent to flame. Buy a good fuse holder, and use a ⁶⁄₁₀-A slow-blow fuse.

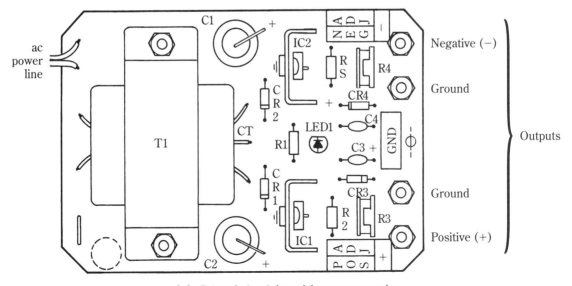

3-3 Printed circuit board for power supply.

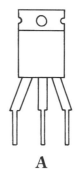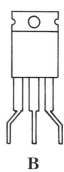

3-4 A. The way some say to bend the leads of the IC regulator. B. The correct way.

A   **B**

The panel light (IL1) tells you when the power supply is turned on. It is a single assembly containing a neon glow lamp and a series current-limiting resistor. I used a 105-Vac to 125-Vac Linrose type B2150A1 that was purchased on a blister-pack display at a local electronic parts distributor. These devices are cylindrical in shape, and are attached to a metal front panel or box (as the case may be) using a "Tinnerman speed nut" (supplied).

Connecting the ac power line is always a matter of concern for any power-supply project. Please understand one thing if nothing else: *Alternating current (ac) from the wall socket is terribly dangerous—it will kill you if you are careless*! The Jameco JE-215 kit comes with a two-wire ac power cord, but I don't recommend its use. Obtain a three-wire ac power cord and a matching chassis-mounted socket (SOC1). There are very potent safety reasons for using the three-wire cord.

Several other parts are added outboard to the printed circuit board. These include output binding posts (BP1 through BP4), an optional metering circuit, and a pair of special output jacks. The binding posts (Fig. 3-5) are sometimes called *five-way binding posts*, after the supposed number of ways you can make connection to them. They provide an insulated path for the dc power through the metal cabinet. In addition, binding posts accept a banana plug, which is a convenient, if temporary, way to carry power from the power supply to your project or experiment on the bench.

Constructing the project

The dual-polarity dc power supply should be built inside a metal cabinet. You can use a decorative cabinet such as I used or a plain "battleship gray" aluminum flange box, whichever you please. The decorative cabinet looks a lot better. I used a Hammond 1458D4B-1184 cabinet, which is 8 × 8 × 4 inches, and painted light blue with a white front panel and black rear panel (Fig. 3-6). There are any number of cabinets on the market, and you can find a good selection at most local or mail-order electronic parts distributors.

Assembly of the printed circuit board follows a straightforward path. Use a pencil-type soldering iron and either 60/40 or 50/50 *resin core* solder (sometimes marked *radio-tv*, *electronic*, or something similar). **Do not use acid-core or coreless solders,** which are intended for plumbing use and are not suitable for electronics. Acid-core solders corrode and destroy the printed board and wiring.

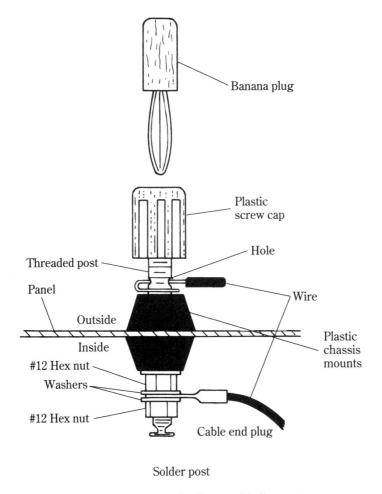

3-5 Connection to the five-way binding post.

Small components such as resistors and rectifier diodes are mounted by passing the leads through the hole in the printed circuit board. Press the body of the part gently against the printed circuit board, causing it to seat firmly, then turn the board over (holding the part in place) and bend over the leads. Make sure to position the leads so that they don't short-circuit to nearby conductors of the printed circuit board. Clip off the excess, leaving ³⁄₁₆ - to ⅛ - inch of lead, and solder. Hold larger components, such as the potentiometers (R3 and R4), in place with your fingers while soldering on the other side of the printed circuit board.

At this point, you have to decide whether you want the power supply to be a fixed-output (i.e., only one output voltage available from each side) or a true variable-output type. If you are satisfied with having only one output voltage (e.g., ±12 V), then mount the potentiometers supplied with the kit at J3 and J4 on the printed cir-

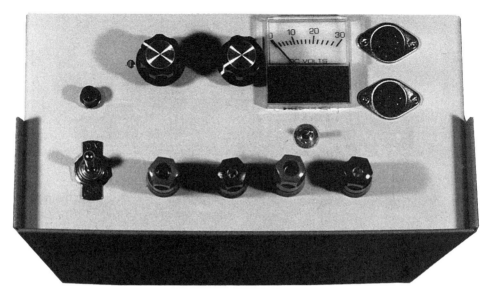

3-6 Front panel of the power supply.

cuit board. If you want the supply to be variable, then discard the potentiometers supplied with the kit and use a pair of potentiometers mounted on the front panel instead. If you opt for the latter, don't install the original potentiometers. Instead install two wires at each potentiometer spot on the printed circuit board. Three holes are provided in the board for each potentiometer; use the two outermost holes in each case.

Mount the printed circuit board approximately in the center of the bottom plate of the Hammond cabinet. Drill three ⁵⁄₃₂ holes for #6 machine screws (mounting hardware) in the bottom plate. These three screws correspond to the three holes in the printed circuit board, two of which are also used to mount the power transformer (T1). Mark these holes on the bottom panel prior to assembling the printed circuit board—it's simply easier. Use either nylon or metal standoffs at each mounting point (Fig. 3-7). These standoffs are cylindrical devices that are either threaded for screws or hollow all the way through to accept a machine screw (the latter are preferred). Use a standoff designed for #6 machine-screw hardware, and is ¼ to ⅜ inch long.

If you cannot find standoffs, or if the offered standoffs are only sold in a large-quantity package, a substitute strategy is possible. I sometimes use two or three hex nuts stacked on top of each other as a de facto standoff (Fig. 3-8). If you are very precise about drilling the mounting holes in the bottom plate, then it is possible to use #6 hex nuts for the standoff. However, if you are less precise, then use the next larger-size hex nuts as the standoffs so a little "slop" will make up for the tolerance problem when mounting the board. For example, in my power supply I used #6 machine screws and hex nuts to secure the printed circuit board and #8 hex nuts for the standoff function.

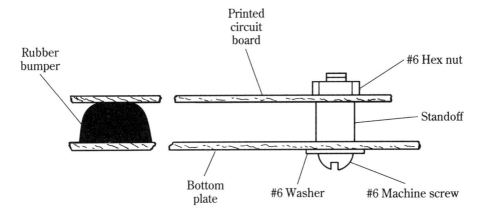

3-7 Rubber bumper used to stabilize PC board.

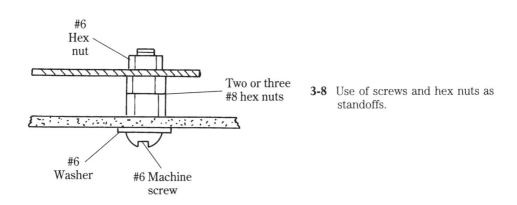

3-8 Use of screws and hex nuts as standoffs.

Jameco supplies an adhesive-backed rubber bumper with the kit. It is placed between the printed circuit board and the bottom plate (Fig. 3-7) at a point indicated in the Jameco instructions. It adds support to the heavy power transformer to prevent warping of the printed circuit board over time.

The rear panel is shown in Fig. 3-9. The ac electrical power connector (SOC1) and fuse holder are mounted on this panel. The connector (SOC1) brings ac from the power-line wall socket safely inside the cabinet. Some people like to use a simple grommet and strain-relief knot to do this job, but there are safety problems with that approach if the power line is strained or the grommet (or panel) is damaged. Use an Underwriter's Laboratory (UL) approved socket. I selected the heavy-duty square type similar to those used on the back of IBM PC computers and most of the clones. These sockets are easily available at electronic parts distributors. Don't forget to buy the matching ac power cord, or you'll wind up robbing the family personal computer when you want to use the power supply.

Mount the fuse holder (F1) on the rear panel alongside the ac power connector (SOC1). Install the 5⁄10-A slow-blow fuse in the fuse holder after you finish soldering the connections, or else some damage may occur to the fuse. The pins on the back of SOC1 will be marked H (hot), N (neutral), and G (ground). Connect the fuse wire to the hot (H) terminal.

Connect the ground terminal of SOC1 to a ground screw that passes through the rear panel (Fig. 3-9 inset). A 5/32-inch hole is drilled in the rear panel at a convenient point close to SOC1. Scrape away the paint on the inside of the panel around the hole until only bright, shiny metal shows. Use a 3/8-inch or longer #6 machine screw and matching hex nut for the ground screw. Make the electrical connection with a ground lug beneath the hex nut. Do not use the smooth form of ground lug, but rather the "star washer" type that bites into the exposed metal. Make two connections to the ground terminal of SOC1, one directly to the rear panel, the other routed to the bottom panel and front panel. That way, all three metal surfaces that are used are grounded as a safety measure against catastrophic ac failure.

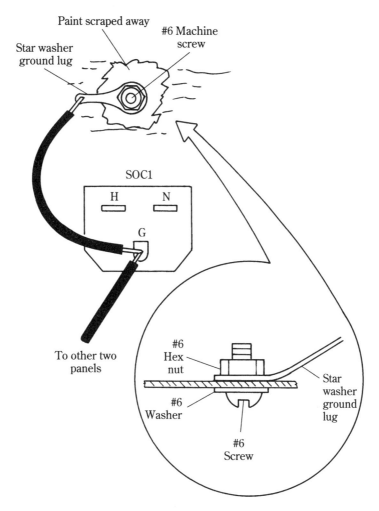

3-9 115 Vac socket wiring. *Danger!* Ac present on this connector can kill you if you touch it. Disconnect the power supply from the ac wall output before working on this circuit.

The internal side of the front panel is shown in Fig. 3-10 (the exterior side was shown previously in Fig. 3-6). The front panel contains the controls, switches, meter, and all output terminals. In the prototype of this project, the meter was an afterthought, so I left too little room for it on the front panel and it is a little crowded. The meter selected is small and is able to read dc voltages up to 30 V. Although any available dc voltmeter of appropriate range (20 V or higher) will do, I found the Modutec 1063-OMS-DVV-030 fit nicely into the limited space that was available.

3-10 Back of front panel showing wiring.

If you build the adjustable version of the power supply, then a meter on the output line is advisable. Some people might want to use two meters, one each for both positive and negative outputs. Indeed, that is the preferred approach if money is no object. But those meters cost from $10 up (with most appropriate types being in the $17 range), so you might want to consider a circuit such as shown as the optional meter circuit in Fig. 3-1, and also in Fig. 3-11. In this circuit, a bridge circuit consisting of four 1N4148 (or equivalent) diodes functions as an automatic polarity-sensitive steering switch for the meter. When switch S2 is connected to the negative power supply ($V-$), the meter current flows through D3, the meter M1, and diode D1 to ground. Alternatively, when S2 is set to read the positive voltage ($V+$), the meter current flows through diode D2, the meter M1, and diode D4 to ground. The diodes cause a slight inaccuracy in the meter reading, but this is tolerable for most applications, especially at potentials above about ±6 V or so.

The meter board containing the diode bridge can be mounted anywhere, but it turned out to be convenient to build it so that the electrical connections from the meter are used to mount the board. The board is made of perforated circuit board. Make a pair of mounting points by placing loops of #22 solid wire at one end of the board and then soldering them to the meter terminals. In some cases (such as the Modutec meter selected here), a small resistor is packed with the meter. Connect this resistor in series with the meter (see R_m in Fig. 3-11).

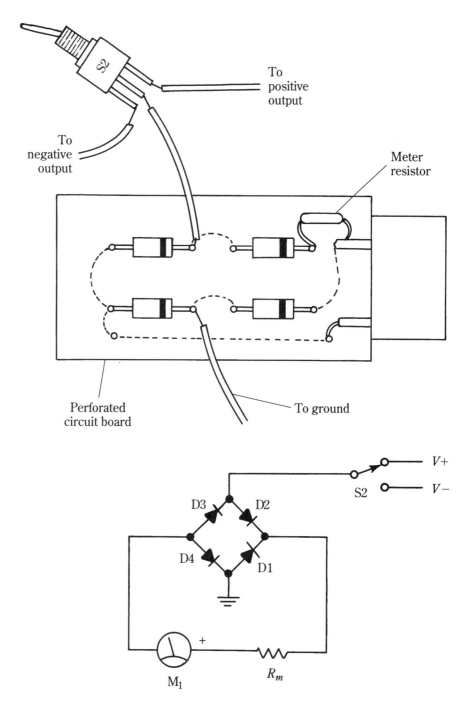

3-11 Meter rectifier circuit.

The special output jacks (J1 and J2) are intended for a series of homebrew instruments and projects that will be built in the future. I did not want to use banana plugs for those applications, and opted instead for a multiconductor dc power cord. I selected a low-cost chassis-mounted 8-pin DIN "audio" connector (Fig. 3-12) because it is low in cost (seemingly simple connectors can cost an arm and a leg!), and is widely available at electronics parts outlets and stores that sell audio supplies.

The overall connection diagram is shown in Fig. 3-13. Shown here are the rear panel, bottom panel, circuit board, and front panel, with appropriate wiring for all parts. Use #14 or #16 stranded wire for wiring on the ac side of the power transformer (T1), and #22 or #24 solid wire for the dc and control functions. In the latter case, use the #22 for the wires from the printed circuit board to the main output binding posts, and the #24 for other wiring.

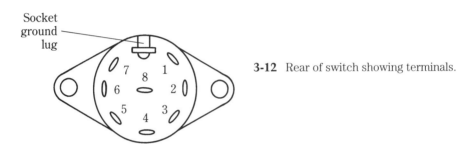

Socket ground lug

3-12 Rear of switch showing terminals.

Adjusting and operating the power supply

One of the nice things about this power-supply project is that it is simple to use. If you opt for the fixed-voltage type, the output voltage must be set adjusting R3 and R4. Connect a dc voltmeter between the positive output terminal (BP1) and ground (BP2). Adjust trimmer potentiometer R3 for the desired output voltage. Next, move the voltmeter to read the potential between BP3 and BP4, and adjust potentiometer R4 for the desired output voltage. When these voltages are adjusted, disconnect the voltmeters and place the top cover on the power supply. Switch S1 turns the power supply on and off.

If you opt for the adjustable-voltage form of this project, then no trimmers are used. The adjustment is made using knobs attached to R3 and R4 on the front panel. Set S2 for either V– or V+, and adjust the associated potentiometer on the front panel (R3 or R4 as appropriate) for the desired output voltage level. Reset S2 to the other polarity, then adjust the remaining potentiometer.

Safety note!

This power-supply project is designed to operate from a 115-Vac power source, such as the wall outlets in your home or shop. The 115-V line can be *fatal* if contacted! Whenever you need to open the power-supply cabinet, it is essential that you unplug the ac power cord both at the wall and at the rear of the power-supply project and physically set it aside. This procedure will prevent accidental electrocution. Don't work on this power supply while it is connected to the ac power line! If you abso-

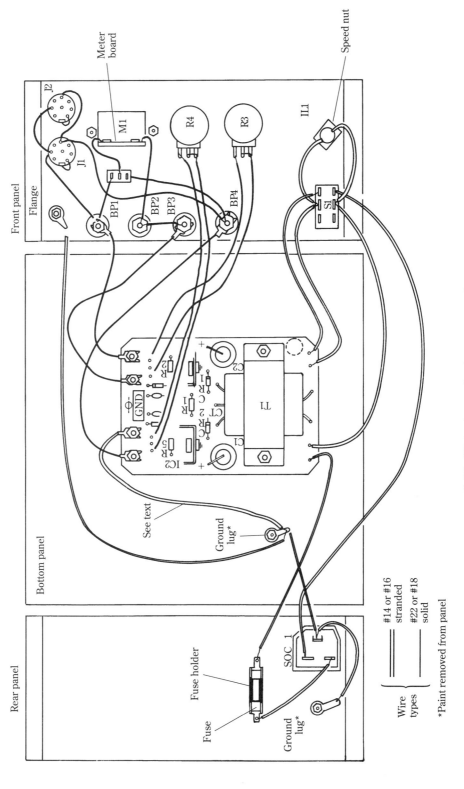

3-13 Wiring layout for dc power supply.

lutely must work on it (as when adjusting the trimmer potentiometers), safety dictates that you use an isolation transformer to power the project. Also, you must be absolutely sure to use plastic adjustment tools, never a metallic screwdriver.

Conclusion

The bipolar dc power supply shown in this article is relatively simple to build and serves as a low-cost solution to the power-supply needs of most electronic experimenters.

PARTS LIST

Jameco JE-215 Power Supply Kit:

IC1	LM-317T IC voltage regulator
IC2	LM-337T IC voltage regulator
C1, C2	2200-µF 16-V electrolytic capacitor
C3, C4	1-µF 35-V tantalum capacitor
CR1–CR4	1N4001 rectifier diodes (1N4001 through 1N4007 can be used)
LED1	Red light-emitting diode
R1	1000-Ω ¼-W resistor
R2, R5	180-Ω ¼-W resistor
R3, R4	2500-Ω trimmer potentiometer (if fixed supply is desired).

Other parts:

R3, R4	2500-Ω 2-W linear taper chassis-mounted potentiometer with ¼-inch shaft (if adjustable version is needed).
S1	GC Electronics 35-132 DPDT switch
IL1	Linrose B2150A1 neon lamp, 105–125 V
S1	SPDT miniature toggle switch
BP1, BP4	Red 5-way binding posts
BP2, BP3	Black 5-way binding posts
SOC1	ac socket, chassis mounted
M1	0- to 20- (or 0- to 30-) Vdc panel meter
J1, J2	8-pin, 270-degree DIN female chassis-mounted connector

Power cord to match ac socket

F1	Fuse holder
	⁶⁄₁₀-A slow-blow fuse to match F1

Augat 92008 p/n PKE5-90B-1.4 knob (two required)

Hammond 1458D4B-1184 8 × 8 × 4 cabinet

Assorted hardware

4
Discrete active devices (transistors)

NEARLY ALL OSCILLATOR AND WAVEFORM-GENERATOR CIRCUITS USE SOME SORT of active amplifying device at their heart. For many circuits, the active device is either a bipolar (npn or pnp) transistor, or a field-effect transistor (FET). Even when the active device is an integrated circuit (IC, see chapter 5) the innards of the device consist largely of transistors. In this chapter you will find out how transistors work.

npn/pnp bipolar pn junction transistors

Although the transistor was not actually invented until the late 1940s, it was predicted by physicists as theoretically possible about 10 years before that time. Natural semiconductor diodes were known as early as pre-World War I when a lead-based mineral crystal called *galena* was used as the detector in radio crystal sets. The semiconductor pn-junction diode is a "blood relative" and precursor to the transistor, but it took metallurgy that grew out of World War II research efforts to stimulate the development of reliable, manufactured semiconductor diodes. Previously, it was not possible to obtain germanium or silicon pure enough to make these devices; intensive wartime research solved that problem. Once that hurdle was out of the way, scientists at Bell Laboratories were able to make the world's first working point-contact transistor. Today, transistors are made in the same variety of ways as are diodes.

Simple bipolar transistors

The word *transistor* is derived from *trans*fer re*sistor*. Transistors are amplifying devices made from semiconductor materials. The simple bipolar transistor is often likened to a pair of diodes connected back to back. The story is much more complex, however, as can be seen if you attempt to obtain anything resembling "transistor action" by connecting a pair of signal diodes.

The basic transistor consists of three sections of semiconductor material, as shown in Figure 4-1A. Two types are shown here, called *npn* and *pnp* (Fig. 4-1B shows their respective circuit symbols). Both consist of a central region of one type of semiconductor material sandwiched between two other regions of the opposite type of material. An npn device, for example, has two n-type regions sandwiching a region of p-type material. In the pnp transistor, just the opposite situation prevails: an n-type central region exists between two p-type sections.

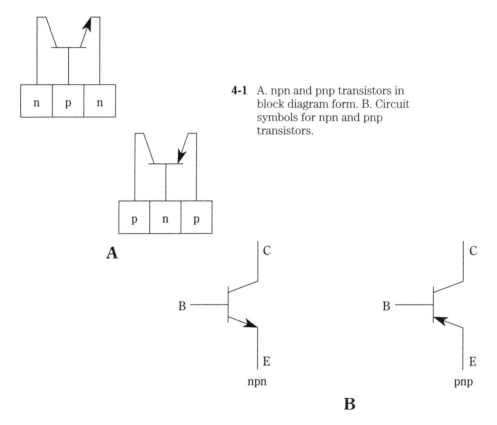

4-1 A. npn and pnp transistors in block diagram form. B. Circuit symbols for npn and pnp transistors.

The central region is called the *base*, and it controls the activities of the entire transistor. The two end sections are called the *emitter* and the *collector*, respectively. You might be tempted to think of these regions as being interchangeable because of the apparent symmetry in the picture. This was true in a very limited sense in the early days of transistors, but today's transistors use a physical geometry that is like our illustration electrically, but is quite different physically. The picture is only a graphical model of the actual situation.

The base region is typically much thinner than either the emitter or collector regions. In addition, the base-region semiconductor material is considerably less heavily doped than either the emitter or base regions. As a result, charge carriers (electrons and holes depending on pnp or npn) crossing the base region have considerably less probability of recombining with the alternate type of charge carrier.

Our block-diagram model of the bipolar transistor is expanded in Fig. 4-2. Here we see an npn transistor, but it also serves as a model for the pnp type as well, provided the V_{EE} and V_{CC} power-supply polarities are reversed.

There are two pn junctions in the bipolar transistor: one formed by the base and emitter (B-E), and the other formed by the collector and base (C-B). In normal operation, the B-E junction is forward-biased and the C-B junction is reverse-biased. At each junction is a barrier potential.

Electrons in the emitter region are repelled by the negative terminal of the V_{EE} potential and attracted into the relatively more positive base region, forming emitter current I_E. Because of the thinness of the base region and its light-doping profile, few electrons recombine with holes. Most of them overcome the C-E junction barrier potential and continue into the collector region. Once in the collector region, the negatively charged electrons come under the influence of the positive terminal of the V_{CC} potential, and are "collected" as current I_C. As current is drawn from the emitter region, additional electrons are attracted into the device from the V_{EE} power supply.

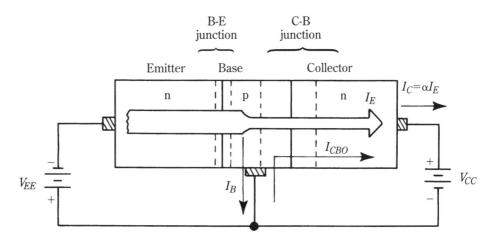

4-2 Charge carrier flow in a bipolar transistor.

This discussion is similar to the pnp, except for the fact that in pnp devices the charge carriers are holes and the battery potentials are reversed. Charge-carrier movement is the same as in Fig. 4-2, although electron flow is reversed.

There is a definite ratio among the emitter current (I_E), base current (I_B), and collector current (I_C). In normal operation:

$$I_E = I_B + I_C \qquad\qquad\qquad \textbf{(4-1)}$$

and,

$$I_C > I_B \qquad\qquad\qquad \textbf{(4-2)}$$

Normally, 95 to 99 percent of the emitter current flows in the collector circuit (i.e., $0.95I_E \leq I_C \leq 0.99I_E$), and only 1 to 5 percent of the emitter current flows in the base circuit.

The collector and emitter currents are different by the amount of the base current, as indicated by Eq. 4-1, and their ratio (I_C/I_E) is called the *alpha* (α) factor. The alpha factor is one way to express transistor gain.

Transistor gain

There are actually several popular ways to denote transistor current gain, but only two are of interest to us here: alpha (α) and beta (β). *Alpha gain* (α) can be defined as the ratio of collector current to emitter current:

$$\alpha = \frac{I_C}{I_E} \tag{4-3}$$

Where: α is the alpha gain
I_C is the collector current
I_E is the emitter current

Alpha has a value less than unity (1), with values between 0.7 and 0.99 being the typical range.

The other representation of transistor gain, and the one that seems more often favored over the others, is the *beta gain* (β), which is defined as the ratio of collector current to base current:

$$\beta = \frac{I_C}{I_B} \tag{4-4}$$

Where: β is the beta gain
I_C is the collector current
I_B is the base current

Alpha (α) and beta (β) are related to each other, and you can use the following equations to compute one when the other is known.

$$\alpha = \frac{\beta}{1 + \beta} \tag{4-5}$$

and,

$$\beta = \frac{\alpha}{1 - \alpha} \tag{4-6}$$

Example

What is the alpha of a transistor that has a beta of 120?

$$\alpha = \frac{\beta}{1 + \beta}$$

$$= \frac{120}{1 + 120}$$

$$= \frac{120}{121}$$

$$= 0.99$$

The preceding values are for static dc situations.

In ac terms you will see *ac alpha gain* (h_{fb}) defined as.

$$h_{fb} = \frac{\Delta I_C}{\Delta I_E} \qquad \text{(4-7)}$$

and *ac beta gain* (h_{fe}) is defined as

$$h_{fe} = \frac{\Delta I_C}{\Delta I_B} \qquad \text{(4-8)}$$

In both equations, the Greek letter delta (Δ) indicates a small change in the parameter with which it is associated. Thus, the term ΔI_C denotes a small change in collector current I_C.

Leakage currents

Another current flowing in the transistor of Fig. 4-2 is the collector-to-base leakage current. Recall from the earlier discussion of pn junctions that the *leakage current* is the movement of minority charge carriers across the pn junction under the influence of the barrier potential. This leakage is designated I_{CBO} in Fig. 4-2.

A note about notation is in order at this point. Several specified currents in the transistor are measured with one of the three terminals (C, B, and E) open-circuited. The basic notation is I_{CBE} with the letter for the open-circuit terminal replaced with an O. Thus, I_{CBO} is the collector-to-base current as measured with the emitter open circuits.

Classifying by common element

Classifying amplifier circuits by common element revolves around noting which element (collector, base, or emitter) is common to both input and output circuits. Although technically incorrect, this is sometimes referred to as the *grounded* element, as in grounded emitted amplifier. I tend to use *common* and *grounded* interchangeably, so bear with me if you are a purist. Figure 4-3 shows the different entries into this class.

Common-emitter circuits

The circuit shown in Fig. 4-3A is the common-emitter circuit. It gets its name from the fact that the emitter terminal of the transistor is common to both input and output circuits. The input signal is applied to the transistor between the base and emitter terminals, while the output signal is taken across the collector and emitter terminals; i.e., the emitter is common to both input and output circuits.

The common-emitter circuit offers high current amplification—the beta rating of the transistor. But the common-emitter circuit also offers a substantial amount of voltage gain as well. The transistor is also a voltage amplifier, especially when a series resistor is placed between the collector terminal and the collector dc power supply. The values of gain for current and voltage are vastly different, however. The current gain is H_{fe}, but the voltage gain depends on other factors as well. Later, you will see that voltage gain depends on the R_L/R_E ratio in some circuits, and the product of that ratio and the beta in other cases.

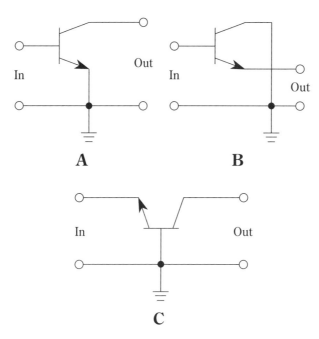

4-3 Transistor amplifier configurations: A. Common emitter.
B. Common collector. C. Common base.

The input impedance of the common-emitter amplifier is medium ranged, or in the 1000-Ω range. The output impedance is typically higher, up to 50 kΩ, however. Typical values are determined by the specific type of circuit, but some approximations can be made. For most common-emitter amplifiers, Z_{in} is equal to the product of the emitter resistor R_E and the H_{fe} of the transistor. The output impedance is essentially the value of the collector load resistor and ranges from 5 kΩ to about 50 kΩ.

The output signal in the common-emitter circuit is 180° out of phase with the input signal. This phase shift means the common-emitter amplifier is an *inverter* circuit. The output signal will be negative going for a positive-going input signal, and vice versa. The common-emitter transistor amplifier is probably the most often used circuit configuration.

Common-collector circuits

The common-collector circuit configuration is shown in Fig. 4-3B. In this circuit, the collector terminal of the transistor is common to both input and output circuits. This circuit is also sometimes called the *emitter-follower* circuit. The common-collector circuit offers little or no voltage gain. Most of the time the voltage gain is actually less than unity (1), but the current gain is considerably higher ($\approx H_{fe} + 1$).

There is no phase inversion between input and output in the emitter-follower circuit. The output voltage is in phase with the input signal voltage.

The input impedance of this circuit tends to be high, sometimes greater than 100 kΩ at frequencies less than 100 kHz. But the output impedance is very low, because it is limited to the value of the emitter resistor, which can be as low as 100 Ω.

This situation leads us to one of the primary applications of the emitter follower: *impedance transformation*. The circuit is often used to connect a high-impedance source to an amplifier with a low-impedance amplifier.

The emitter follower, *née* common-collector amplifier, is also frequently used as a *buffer amplifier*, which is an intermediate stage used to isolate two circuits from each other. One primary example of this application is in the output circuit of oscillator circuits. Many oscillators will "pull," or change frequency, if the load impedance changes. Yet some of the very circuits used with oscillators naturally provide a changing impedance situation. The oscillator proves a lot more stable under these conditions if an emitter-follower buffer amplifier is used between its output and its load.

Common-base circuits

Common-base amplifiers use the base terminal of the transistor as the common element between input and output circuits (Fig. 4-3C). The output is taken between the collector and base.

The voltage gain of the common-base circuit is high, on the order of 100 or more; however, the current gain is low, usually less than unity. The input impedance is also low, usually less than 1000 Ω, because it is limited to the emitter resistance. On the other hand, the output impedance is quite high. Again, there is no phase inversion between input and output circuits.

The principal use of the common-base circuit is in high-frequency (HF) and very high frequency (VHF) RF amplifiers in receivers. The circuit requires no neutralization at these frequencies, so it is superior to common-emitter circuits. Neutralization prevents oscillation because of interelement capacitances that provide a feedback signal in phase with the input signal.

dc load lines, the "Q" point, and transistor biasing

The dc load line is a graphical method for expressing transistor operating conditions. The basis for the load line is the I_C-vs-V_{CE} family of curves (Fig. 4-4A). The curve is traced for a circuit similar to Fig. 4-4B at several different base currents, as the collector voltage is swept from zero to maximum.

Figure 4-4B shows a common-emitter npn transistor circuit. The supply voltage is 30 Vdc; the collector current I_C can vary from zero to a maximum of 12 mA (0.012 A). A 2500-Ω collector load resistor (R_L) connects the collector of the transistor to the V_{CC} power supply. The collector-emitter voltage (V_{CE}) can rise to V_{CC} when $I_C= 0$, or drop to zero when I_C is maximum. The collector current is set by the base current, I_B, hence also the base voltage V_B. At any given point, the value of V_{CE} is V_{CC} less the voltage drop across the load resistor (V_R). Because V_R is the product I_CR_L,

$$V_{CE} = V_{CC} - I_CR_L \qquad (4\text{-}9)$$

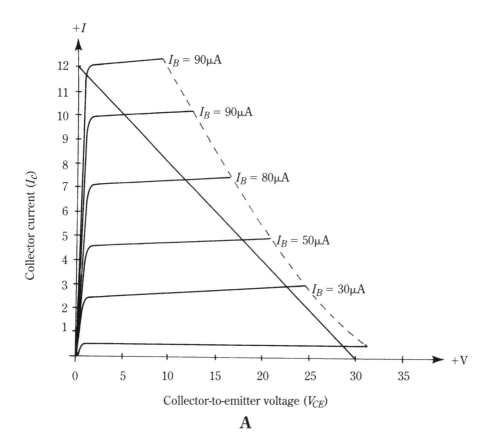

A

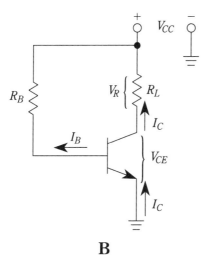

B

4-4 A. Family of operating curves and dc load lines. B. Common emitter npn transistor circuit from which family of curves (A) was traced.

Example

For Fig. 4-4B, calculate V_{CE} when I_C = 7 mA.

Solution:

$$V_{CE} = V_{CC} - I_C R_L$$
$$= (30 \text{ V}) - ((0.007 \text{ A}) (2500 \text{ } \Omega))$$
$$= (30 \text{ V}) - (17.5 \text{ V})$$
$$= 12.5 \text{ V}$$

Transistor biasing

Biasing sets the operating characteristics of any particular transistor circuit and is usually set by the current conditions at the base terminal of the device. Several different bias networks are commonly seen in transistor circuits, and they are summarized here.

Fixed-current bias   A fixed-current bias circuit (Fig. 4-5A) sets a base current in the transistor at a fixed resistor between the V_{CC} power supply and the base terminal of the transistor. That current, I_B, is defined by:

$$I_B = \frac{V_{CC} - V_{BE}}{R_B} \qquad \text{(4-10)}$$

To see how values for components are reached in a circuit such as Fig. 4-5A, let's work an example.

Example

A silicon npn transistor has a current beta gain H_{fe} of 90, and is to be operated from a 12-Vdc power supply, with a no-signal quiescent Q-point V_{CE} potential of 6 V and a collector current of 2 mA (0.002 A).

1. Write down what we know:

   V_{CC} = 12 Vdc

   V_{CE} = 6 Vdc

   I_C = 0.002 A

   V_{BE} = 0.6 V (Si transistor is being used)

2. Calculate the value of load resistor R_L by Ohm's law:

$$R_L = \frac{V_{CC} - V_{CE}}{I_C}$$
$$= \frac{(12) - (6)}{0.002 \text{ A}}$$

$$= \frac{6}{0.002}$$
$$= 3000 \ \Omega$$

3. Calculate base current I_B. By definition $H_{fe} = I_C/I_B$, so:

$$I_B = \frac{I_C}{H_{fe}}$$
$$= \frac{0.002 \ A}{90}$$
$$= 2.2 \times 10^{-5} \ A$$
$$= 22 \ \mu A$$

4. Calculate the value of base resistor R_B by rearranging Eq. 4-10 to solve for R_B:

$$R_B = \frac{V_{CC} - V_{BE}}{I_B}$$
$$= \frac{(12) - (0.6)}{2.2 \times 10^{-5} \ A}$$
$$= \frac{11 \ V}{2.2 \times 10^{-5} \ A}$$
$$= 518{,}182 \ \Omega$$

As a practical matter, standard resistor values are used, so R_L would be selected as either 2700 Ω or 3300 Ω, and R_B as 510 kΩ.

The circuit of Fig. 4-5A presents several problems, including a dependence on the transistor beta gain (H_{fe}) and the value of V_{BE}. Variations in actual—versus published "typical"—H_{fe} is approximately 0.55 to 0.7 V in silicon transistors (0.2 to 0.3 V in germanium devices), and this barrier potential is a function of temperature. A variant of the fixed bias circuit, shown in Fig. 4-5B, helps solve some of these problems.

The principal difference between Figs. 4-5A and 4-5B is the use of an emitter resistor, R_E. The voltage drop across this resistor must be added to V_{BE} when calculating R_B, and is:

$$V_E = I_E R_E$$
$$0 = (I_C - I_B) R_E \qquad\qquad \textbf{(4-11)}$$

or:

$$V_E = \left(I_C - \frac{I_C}{H_{fe}} \right) R_E \qquad\qquad \textbf{(4-12)}$$

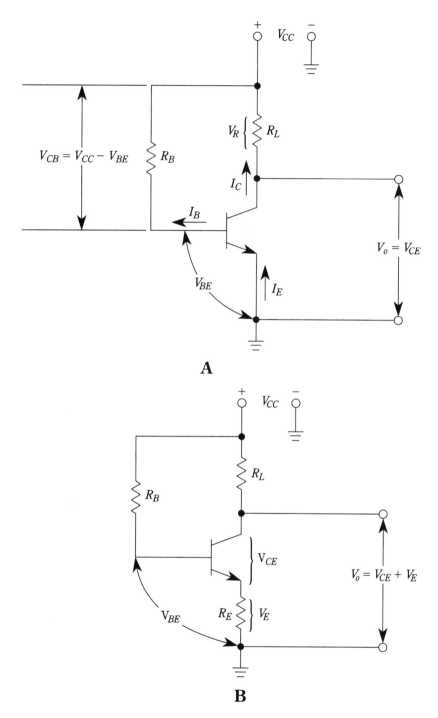

4-5 A. Voltage and current relationships in biased transistor. B. Circuit modified for emitter resistor.

A rule-of-thumb for R_E is that it should be approximately one-tenth of R_L, and in no case be more than one-fifth R_L; most of the time $10 \, \Omega \leq R_E \leq 1000 \, \Omega$.

Returning to our example, let's set $R_L = 3{,}000 \, \Omega$, $R_E = 220 \, \Omega$, and $I_C = 0.002$ A. In that case,

$$V_E = (I_C - I_B) \, R_E$$
$$= (0.002 \text{ A} - 0.000022 \text{ A}) \, (220 \, \Omega)$$

$$R_B = \frac{V_{CC} - (V_{BE} + V_E)}{I_B}$$

$$= \frac{(12 \text{ V}) - (0.6 + 0.44 \text{ V})}{2.2 \times 10^{-5}}$$

$$= \frac{(12 \text{ V}) - (1.04)}{2.2 \times 10^{-5} \text{ A}}$$

$$= 498{,}182 \, \Omega$$

For the circuit of Fig. 4-5B, the following relationships obtain:

$$Z_o = R_L$$
$$Z_{in} = R_E H_{fe}$$
$$A_I = H_{fe}$$
$$A_v = R_L H_{fe}/R_E$$

Collector-to-base bias In this type of bias network, the resistor supplying bias current to the base (R_B) is connected to the collector of the transistor, rather than V_{CC} (see Fig. 4-6). An advantage of this circuit is that the quiescent (no-signal) conditions are stabilized somewhat because I_B is set by V_{CE}, rather than V_{CC}. Thus, when I_C tries to increase, the voltage drop across R_L increases, and because $V_{CE} = V_{CC} - V_{RL}$, the value of V_{CE} decreases. This action, in turn, reduces I_B, so, by $I_C = H_{fe}I_B$, the collector current decreases.

A similar action takes place when I_C tries to decrease. The end result in both cases is that I_C tends to stabilize around the quiescent value.

In terms of the previous example, current I_B is set by V_{CE}, rather than V_{CC}, so:

$$V_{CE} = V_{CC} - R_L \, (I_B + I_C) \tag{4-13}$$

and,

$$V_{CE} = I_B R_B + V_{BE} \tag{4-14}$$

which can be rewritten

$$R_E = \frac{V_{CE} - V_{BE}}{I_B} \tag{4-15}$$

$$R_B = \frac{(6 \text{ V}) - (0.6 \text{ V})}{2.2 \times 10^{-5} \text{ A}}$$

$$= \frac{5.4}{2.2 \times 10^{-5}\ \text{A}}$$

$$= 245{,}455\ \Omega$$

As was true in the previous bias circuit, it is sometimes prudent to insert an emitter resistor to gain further stability, as in Fig. 4-6B. For the circuit of Fig. 4-6B:

$$Z_o = R_L$$
$$Z_{in} = R_E H_{fe}$$
$$A_I = H_{fe}$$
$$A_v = R_L H_{fe}/R_E$$

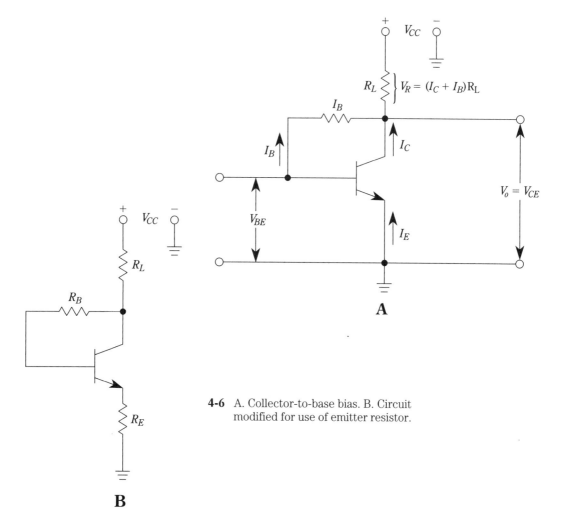

4-6 A. Collector-to-base bias. B. Circuit modified for use of emitter resistor.

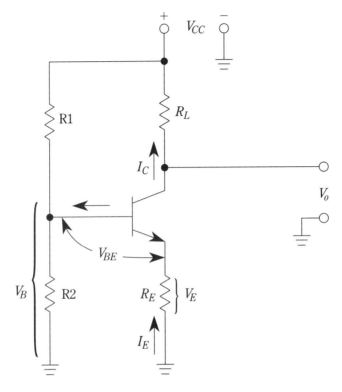

4-7 Voltage divider bias scheme for most stable operation.

Emitter bias or "self-bias"   Figure 4-7 is recognized as the most stable configuration for transistor amplifier stages. This circuit uses a resistor voltage divider
(R_1/R_2) to set a fixed bias voltage (V_B) on the transistor. As a general rule, the best
stability usually occurs when $R_1\|R_2 = R_E$. Because there is a substantial voltage drop
across R_E, the V_{CC} voltage required for Fig. 4-7 is a bit higher than for the previous
circuits. For purposes of this discussion let $V_{CC} = 18$ Vdc, $V_C = 9$ Vdc, and $V_E = 6$ Vdc.
As in the previous examples, let $H_{fe} = 90$ and $I_C = 0.002$ A. The base current is:

$$I_B = \frac{I_C}{H_{fe}} \qquad\qquad\qquad \textbf{(4-16)}$$

$$= \frac{0.002 \text{ A}}{90}$$

$$= 2.2 \times 10^{-5} \text{ A}$$

Load resistance R_L is:

$$R_L = \frac{V_{CC} - V_C}{I_C} \qquad\qquad \textbf{(4-17)}$$

$$I_E = I_C - I_B$$

$$= 0.00198 \text{ A}$$

$$R_E = \frac{V_E}{I_E}$$

$$= \frac{6 \text{ V}}{0.00198 \text{ A}}$$

$$= 3030 \ \Omega$$

$$V_B = V_E + V_{BE} \qquad \text{(4-18)}$$
$$= 6 \text{ V} + 0.6 \text{ V}$$
$$= 6.6 \text{ V}_{dc}$$

$$V_B = \frac{V_{cc}R_2}{R_1 + R_2} \qquad \text{(4-19)}$$

$$\frac{V_B}{V_{cc}} = \frac{R_2}{R_1 + R_2} \qquad \text{(4-20)}$$

$$\frac{6.6 \text{ V}}{18 \text{ V}} = \frac{R_2}{R_1 + R_2}$$

$$= 0.37$$

or, when the arithmetic is done:

$$R_1 = 1.7R_2 \qquad \text{(4-21)}$$

Further, $R_1 \| R_2 = R_E$, so:

$$\frac{R_1 R_2}{R_1 + R_2} = 3030 \ \Omega \qquad \text{(4-22)}$$

Solving Eq. 4-22 for R_2 and then plugging in R_1 shows that R_1 4812 Ω and $R_2 = 8181 \ \Omega$.

Frequency characteristics

Transistors, like most other electron devices, operate only over a certain specified frequency range. Here are three basic cutoff frequencies that may interest us: *alpha*, *beta*, and the *gain-bandwidth product* (f_t).

The alpha cutoff frequency f_{ab} is the frequency at which the ac current gain h_{fb} drops to a level 3 dB below its low-frequency (usually 1000 Hz) gain. This is the frequency at which $h_{fb} = 0.707h_{fbo}$, where h_{fbo} is the ac current gain at 1000 Hz.

The beta cutoff frequency is similarly defined as the frequency where the ac beta, h_{fe}, drops 3 dB relative to its 1000-Hz value. In general, this frequency is lower than the alpha cutoff, but is considered somewhat more representative of a transistor's performance.

The frequency specification that seems to be quoted most often is the gain-bandwidth product, which is given the symbol f_t. This parameter is usually accepted only for transistors operated in the common-emitter configuration. It is defined as:

$$f_t = \text{Gain} \times \text{Bandwidth} \qquad \text{(4-23)}$$

$$f_t = h_{fe}f_o \qquad \text{(4-24)}$$

where: f_t is the gain-bandwidth product
$\qquad$ h_{fe} is the ac beta
$\qquad$ f_o is the frequency at which gain is measured

The value of f_t quoted in specification sheets is the frequency at which h_{fe} drops to unity.

If the beta cutoff frequency f_{ae} is known, then the gain-bandwidth product may be approximated from

$$f_t = f_{ae}h_{feo} \qquad \text{(4-25)}$$

Recognize, however, that this is an approximation that may not hold up in every case. Also, you can often get away with assuming that f_o is approximately equal to, but usually slightly less than, the alpha cutoff frequency.

Transistor packages

The classic small transistor case is the TO-5 package shown in Fig. 4-8. This package was used as early as 1954 and consists of a metal can in which the transistor die is mounted. Plastic transistors soon replaced the metal (although metal is still used in high-reliability and military applications). A similar sized plastic package is sometimes advertised as "similar to TO-5." The TO-92 metal package is even smaller than the TO-5, but contains the same sort of small-signal transistors. Like its TO-5 cousin, the TO-92 inspired plastic versions (Fig. 4-9).

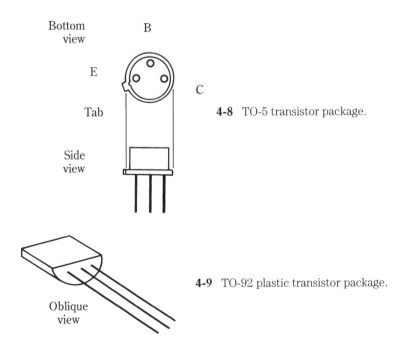

4-8 TO-5 transistor package.

4-9 TO-92 plastic transistor package.

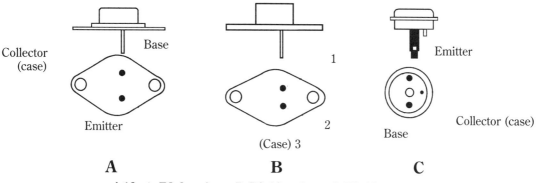

4-10 A. TO-3 package. B. TO-66 package. C. TO-36 package.

Figure 4-10 shows several types of power transistor packages. The TO-3 transistor in Fig. 4-10A is the so-called "standard" power transistor in a diamond-shaped package. A smaller diamond-shaped package is the TO-66, shown in Fig. 4-10B. A Japanese "similar-to-TO-66" package looks at first like the TO-66, but has slightly different pin spacings. Finally, the big package shown in Fig. 4-10C is the TO-7, or TO-36, depending on power level. This high-power transistor is used extensively in automotive audio applications, mobile two-way radio solid-state HV multivibrator dc power supply converters, and in industrial electronics applications. Older tube-type mobile transmitters often used these transistors in the 13.6 Vdc-to-HV power supply.

Figure 4-11 shows several popular plastic power-transistor packages. Some of these are listed as replacements for TO-3 or TO-66 diamond-shaped power transistors. The package in Fig. 4-11A is the TO-220, once also called "P-66," and is common in small, low-to-medium-powered audio applications and most car radios. Two additional tab-mounted plastic power transistors are shown in Figs. 4-11B and 4-11C. Finally, the device shown in Fig. 4-11D is representative of a class of Motorola power transistor. These devices are not tab-mounted, but instead have a mounting hole through the body of the transistor.

Figure 4-12 shows how a plastic, tab-mounted TO-220 power transistor can replace a TO-3 power transistor. The center terminal (collector) is cut off of the TO-220—it won't be needed, the tab mount is also connected to the collector—and the base and emitter leads are bent down. The mounting screw is passed through the tab hole into the original mounting hole for the TO-3 transistor.

Another transistor package is the RF power transistor in Fig. 4-13. This device uses thin, flat "low inductance" leads. Several sizes are available, and size doesn't always indicate relative power dissipation rating, although it usually does.

There was one 150-MHz FM mobile power amplifier that came with either of two different types of power transistor. The hole in the printed wiring board was cut for the larger type, and rubber O-rings were placed around the smaller to make them fit.

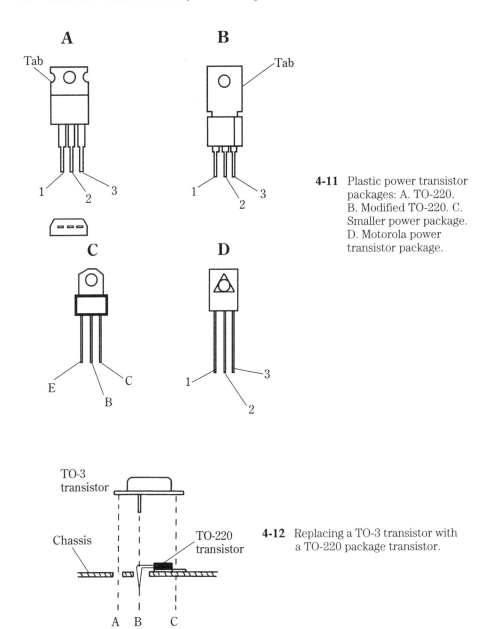

4-11 Plastic power transistor packages: A. TO-220. B. Modified TO-220. C. Smaller power package. D. Motorola power transistor package.

4-12 Replacing a TO-3 transistor with a TO-220 package transistor.

Field-effect transistors

The *field-effect transistor* (FET) was the subject of scientific speculation in the late 1930s. A paper published in a physics journal during that decade gave details for the construction of the *junction field-effect transistor* (JFET), and that hypothetical transistor would have worked if the metallurgy of the era had been better. Following World War II, it was the bipolar transistor that received research attention. If the

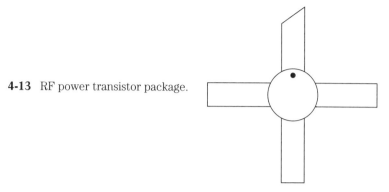

4-13 RF power transistor package.

JFET had been built in 1939, then the semiconductor era would've seen a ten-year head start over what actually occurred—and it's likely that most transistors today would be field-effect devices rather than bipolar devices.

FETs have a very high input impedance, and are transconductance amplifiers. They tend to have performance properties that are far more reminiscent of the pentode vacuum tube than of bipolar transistors.

There are two basic types of field-effect transistor device: the aforementioned *junction field-effect transistor* (JFET) and the *metal-oxide semiconductor field-effect transistor* (MOSFET); the MOSFET is also sometimes called the *insulated-gate field-effect transistor* (IGFET).

Junction field-effect transistors

A basic JFET device is shown in Figs. 4-14 and 4-15. The JFET consists of a semiconductor channel and a gate structure. Two basic types of JFET are seen: *p-channel* and *n-channel*. The designations are based on the type of semiconductor material used to form the channel section. There are three electrode elements on the JFET: *gate, source,* and *drain.*

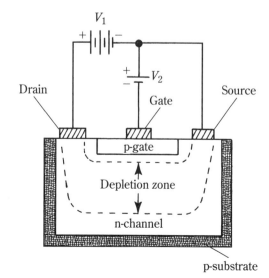

4-14 Junction field-effect transistor (JFET) biased for conduction.

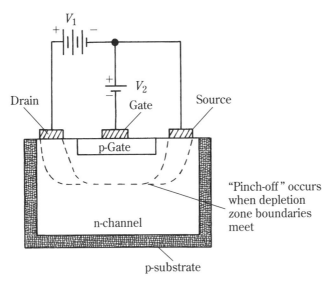

4-15   JFET device in pinch-off.

The gate structure is always made of the opposite type of semiconductor material from the channel structure. In normal operation, a voltage is applied across the channel in the polarity shown. Some JFETs, incidentally, are capable of operation when the drain and source ends of the channel are reversed.

The channel has an electrical resistance less than infinity, so current can flow. When a potential is applied to the gate, however, it will create a *depletion zone* in the channel material. Very few electrical charge carriers (electrons or holes) can exist in this zone, so it has a very high electrical resistance. At 0 V gate potential, the depletion zone is at a minimum, so the channel resistance is the lowest. As the potential is increased, however, the depletion zone widens, thereby narrowing the portion of the channel in which charge carriers will flow. At some specific potential, called the *pinchoff voltage*, the depletion zone completely chokes off the channel and no current can flow from source to drain.

JFET circuits

The circuit of Fig. 4-14 is for an n-channel JFET. The gate structure, made of a different form of semiconductor material than the channel, is diffused into the channel in a manner resembling the emitter of a bipolar transistor. However, this junction is kept perpetually reverse biased. In fact, if the junction were to become forward-biased, destruction of the device is a distinct possibility. The channel is formed such that it is placed between the p-type gate and a p+-type substrate on which the transistor is built.

Electrons in the n-channel are encouraged to flow by the external potential source. If the depletion zone is wide, then current will flow in the external circuit. By making the gate-substrate boundary negative with respect to the channels, electrons are repelled (like charges repel each other) in the channel, thereby widening the depletion zone. When the depletion zones from gate and substrate meet in the channel, pinchoff occurs. This effect is shown in Fig. 4-15.

Consider the channel resistance for a moment. When the gate voltage is 0, the depletion zones are narrow, so the channel resistance is low. When a high negative potential is applied to the gate and substrate, however, the depletion zones widen, causing the channel resistance to increase. When the gate potential reaches a certain critical level, the pinch-off voltage, the channel resistance becomes extremely high. The JFET, then, can be used as an electronic switch because it possesses an extremely high channel resistance under one voltage condition, and a very low resistance under another. The CMOS digital device, type number 4066, is a series of six FETs used specifically as switches.

The p-channel device is exactly the same as the n-channel JFET, except for polarity. The channel structure will be made of p-channel semiconductor material, while the gate and substrate will be made from n-channel material. The circuits for p-channel devices are the same as for n-channel circuits, but with the external voltage polarities reversed. Symbols for the two types of device are shown in Fig. 4-16.

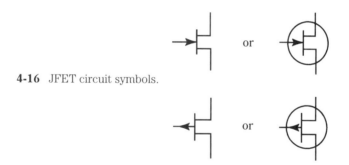

4-16 JFET circuit symbols.

MOSFET devices

MOSFET devices are a bit different from JFET devices. MOSFETs are capable of much higher input impedances than JFETs because there is no physical connection (i.e., an emitter-like pn junction) between the gate and the channel. In the JFET device, input (i.e., *gate*) impedance is limited by the reverse leakage current across the reverse-biased gate-channel junction.

There are two basic types of MOSFET device: *depletion* and *enhancement*. These devices operate in somewhat different modes, and will therefore be discussed separately. Both types of MOSFET device are available in both p-channel and n-channel versions. Thus, the possible simple MOSFETs are:

- p-channel depletion MOSFET
- p-channel enhancement MOSFET
- n-channel depletion MOSFET
- n-channel enhancement MOSFET

MOSFET devices are sometimes called *insulated-gate field-effect transistors* (IGFET) because of their gate structure.

Depletion MOSFETs

The depletion type of MOSFET is shown schematically in Fig. 4-17. As in the JFET device, the channel will be of one of the two basic types of semiconductor material— n or p—and the substrate will be of the opposite type of material. It is always desirable to keep the substrate-channel pn junction either reverse-biased or zero-biased to prevent an excessive current flow. The gate is a metallic contact insulated from the channel material by an extremely thin metal-oxide insulating material. This is the origin of the name, *insulated gate*. The depletion MOSFET is normally *biased on*, meaning that the channel resistance is low when the gate voltage is zero.

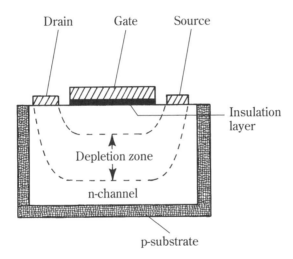

4-17 Depletion-mode metal-oxide semiconductor field-effect transistor (MOSFET).

When an electrical potential is applied to the gate, however, an electrical field is created in the channel. If the voltage is positive with respect to the channel, electrons will tend to draw closer to the gate structure. If the voltage is negative, however, the field will repel electrons in the channel, creating a depletion zone. The higher the gate voltage, the wider the depletion zone. When the depletion zone is finally wide enough to pinch off the channel, current flow ceases. We can, therefore, control the channel resistance with a gate voltage.

Because channel resistance is controlled by the gate voltage, the output current flowing in the source-drain circuit is also under the control of the gate voltage (assuming constant voltage between drain and source).

Field-effect transistors are *transconductance amplifiers*. The definition of such amplifiers is that a change of input voltage (ΔV_{in}) causes a change of output current (ΔI_o). In the transfer-function format that takes the ratio of output to input:

$$g_m = \frac{\Delta I_O}{\Delta V_{in}} \tag{4-26}$$

The unit of transconductance is the *siemen* (S); one siemen equals one ampere per volt; i.e., 1 S = 1 A/1 V. Formerly the siemen was called the *mho* to recognize that it is the inverse of resistance ("mho" is "ohm" spelled backwards). For practical FET devices, the unit *millisiemen* ($\frac{1}{1000}$ of a siemen) is typically used to express the gain of the device.

The symbol for a depletion type MOSFET is shown in Fig. 4-18. As in the case of JFET devices, the arrow pointing inward denotes an n-channel device, while an arrow pointing outward denotes a p-channel device. Incidentally, the manufacture of p-channel depletion transistors is somewhat more difficult than the manufacture of n-channel devices. As a result, n-channel devices predominate in the marketplace.

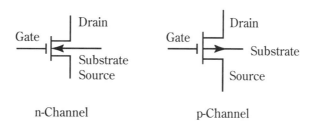

n-Channel   p-Channel

4-18 Depletion mode MOSFET circuit symbols.

Enhancement MOSFETs

The enhancement MOSFET is a normally off device (exactly the opposite of the depletion-type MOSFET) in which it is necessary to apply a gate potential before channel conduction takes place. Again, both p-channel and n-channel devices are possible. The p-channel enhancement MOSFET requires a negative gate potential to begin conduction, while the n-channel device requires a positive gate potential.

Figure 4-19 shows the construction of the enhancement MOSFET. In this case, an n-channel device is shown.

In all MOSFETs is an inherent tendency for electrons to cluster close to the interface between the metal-oxide layer and the channel material. This phenomenon forms a thin n-type region in the p-type substrate, located immediately beneath the insulating layer. The drain and source regions are n-type semiconductor material diffused into the p-type channel material. Ordinarily, no conduction occurs between them, except for the tiny n-type region close to the gate structure. When the gate voltage is zero, the electrons of the n-channel are prevented from migrating out of the drain and source regions. However, when a positive potential is applied (Fig. 4-20) to the gate, the electrons from these regions are attracted to the insulator-substrate interface, causing a conduction zone to appear between the source and drain. Current will flow under this circumstance.

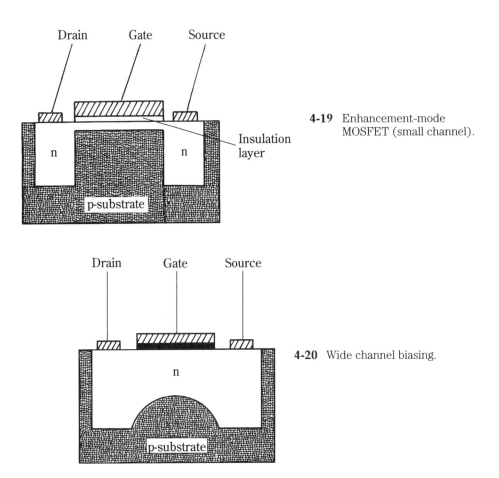

4-19 Enhancement-mode MOSFET (small channel).

4-20 Wide channel biasing.

The conduction of the enhancement MOSFET is controlled by varying the gate voltage. This action will increase or decrease the attraction of the drain and source electrons toward the gate region.

The circuit symbols for the enhancement-type MOSFET are shown in Fig. 4-21. These symbols are the same as the depletion MOSFET symbols, except that the channel bar is broken into sections to indicate the enhancement action taking place inside the device.

In both types of MOSFETs, the activity can be used as an electronic switch. In both cases, one condition causes a low channel resistance, while the opposite condition causes an extremely high "pinched off" resistance. The on-resistance depends, in part, on the signal voltage applied, but it is typically between 10 Ω and 2000 Ω. Most devices will have a resistance in the 100-to-200-Ω range.

The variation of on-resistance is something of a problem that can lead to distortion of the output signal, but manufacturers have been able to limit this problem by using a load resistance that is high compared with the channel resistance. Load resistances on the order of 10 kΩ to 500 kΩ are often seen.

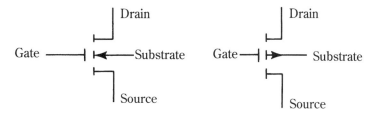

4-21 Enhancement-mode MOSFET circuit symbols.

The off-resistance of the channel is usually many megohms. The specification is usually given in terms of leakage current, rather than resistance. In most switches, the leakage current under the off condition, called the *effective off resistance*, will be from 0.01 to 100 pA. Assuming a +10 V applied signal, then, and a "typical" 10 pA leakage current, the resistance would be:

$$R_{off} = \frac{10\text{ V}}{10^{-8}\text{ A}} = 1000\text{ M}\Omega \tag{4-27}$$

The on resistance of the channel, on the other hand, drops to a much lower value, such as 100 Ω or so. This fact, incidentally, makes the FET almost ideal as an analog electronic switch. A control voltage can be applied to the gate in order to obtain an expected switching action from the difference between the on resistance and off resistance of the channel.

The primary use of the field-effect transistor is as an amplifier, or as the amplifying element in an oscillator or active filter circuit. The input impedance of the FET is typically very high. For some JFETs on the low end of the price scale, this figure may be less than 1 MΩ, but for most it is quite a bit higher. In some MOSFET devices, the input impedance exceeds *one terraohm* (10^{12} Ω).

In some circuit configurations, the high input impedance, coupled with a generally low output impedance, makes the device useful as an impedance converter. An amplifier for a signal source that has a high internal impedance can be made by using a field-effect transistor in the "front-end" circuit.

Later sections will consider some aspects of FET circuit design and the types of circuit that might be selected. In Fig. 4-22, however, one elementary FET amplifier circuit is demonstrated.

The circuit of Fig. 4-22 shows a *common-source amplifier,* which is analogous to the common-emitter circuit seen earlier in the discussion of bipolar transistors. The input impedance of the common-source amplifier is essentially the value of R_1; the JFET input resistance is usually much higher than this resistor. The input signal voltage is applied across the gate and source terminals of the JFET. The source terminal must be bypassed for ac by capacitor C3. The value of this capacitor must follow the one-tenth rule: the capacitive reactance of C3 must be less than, or equal to, one-tenth the value of R_3 at the lowest operating frequency.

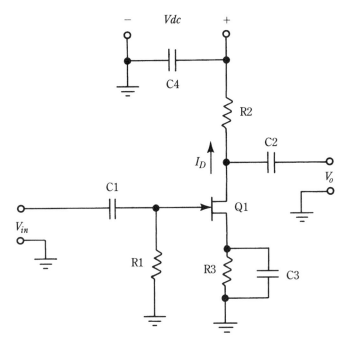

4-22 JFET amplifier circuit with self-bias.

The amplifier can operate as a voltage amplifier through a mechanism that is very similar to the scheme used to make a bipolar npn or pnp transistor operate as a voltage amplifier. In the case of the FET (a transconductance amplifier device), we use a resistance in series with the drain and the dc power supply. The channel of the FET operates as an electronically variable resistance when a signal voltage is applied to the gate. When the signal voltage increases in a positive direction, the channel resistance drops, so the voltage at the drain is reduced (the percentage of the terminal voltage dropped across R2 increases). Similarly, when the drain resistance increases, the voltage at the drain also increases. If the bias on the FET is set at a point where the drain voltage is $(V+)/2$, the output voltage will swing positively and negatively about this potential.

The bias is set by the voltage drop across resistor R3, which is analogous to self-bias in bipolar circuits. This voltage is equal to the product of the source-drain current and the resistance of R3. We normally want the gate to be negative with respect to the source (this is an n-channel device), so making the source more positive than the gate serves the same purpose.

Operating regions

There are two regions of operation for the field-effect transistor. In the *ohmic region*, which exists for low drain-source voltages (V_{DS}), the channel current is directly proportional to the voltage. In the *pinch-off region*, which is also called the *constant current region*, the current will not increase for further increase in V_{DS}. It is this latter characteristic that allows the FET to operate as a constant-current source (CCS). In this application, the source and gate are tied together, and the device operates as a diode with constant-current characteristics. Several CCS devices available on the market are nothing more than internally connected JFETs.

A field-effect transistor can be biased in any of several different ways. Self-bias can be used, as can external bias and certain combinations of self and external bias. Figure 4-22 shows two methods of FET biasing. Figure 4-23A shows an example of external bias. A reverse-bias dc potential is applied to the gate of the FET through resistor R1. The resistor limits current flow across the junction and provides a load for the input signal (if needed). It also allows the discharge of electrons that build up on the gate structure. The level of bias is set by varying *V1*.

An example of self-bias, which is by far the most common form, is shown in Fig. 4-23B. In this case, the voltage drop across source resistor R2 is the bias voltage. The bias must make the gate more negative (or less positive) than the source. A negative voltage was applied in the previous example, but in this circuit the source is made more positive than the gate (which is conceptually the same thing).

The value of the resistor must take into consideration the source-drain current of the FET and the desired level of bias. Ohm's law is applied. We know that the gate-source voltage is given by:

$$V_{GS} = -I_D R_S \qquad \textbf{(4-28)}$$

Where: V_{GS} is the gate-source voltage
I_D is the drain-source current
R_S is the source resistor

A power supply potential must be supplied to the FET (i.e., V_{DC}), but only a portion of the power supply voltage appears across the drain-source (V_{DS}). The voltage drops around the circuit are:

$$V_{DD} = I_D(R_D + R_S) + V_{DS} \qquad \textbf{(4-29)}$$

Where: V_{DD} is the dc power supply voltage
I_D is the drain-source current
R_S is the source resistor
R_D is the drain resistor
V_{DS} is the drain-source voltage

We could select a value for the collector current (see the data sheet for any specific device) and a drain-source voltage. In most cases, the value of V_{DS} will be one-half of V_{DD}. For example, if we applied 28 Vdc to the circuit (V_{DD}), and want a 2-mA current to flow in the drain circuit, we would select values for the resistors consistent with this goal. In general, we will make the source resistor one-tenth to one-fifth the drain resistor.

Graphical method

One of the methods for determining the proper values for the source resistor is to use the graph of $I_D - VS - V_{GS}$ characteristic of the particular device under consideration. These graphs are printed in the manufacturer's data sheets. An example is shown in Fig. 4-24.

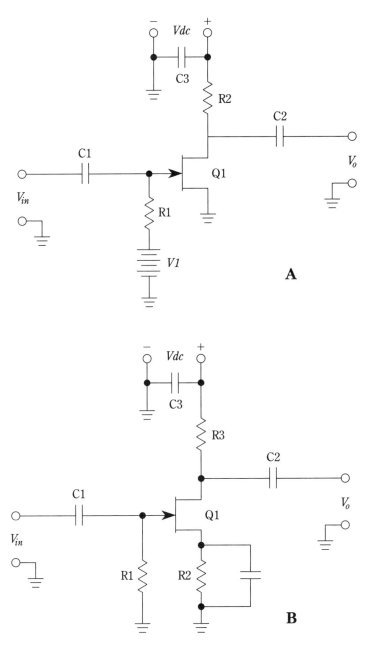

4-23   A. JFET amplifier with external bias. B. Practical JFET amplifier circuit.

A load line is drawn from the pinch-off voltage region on the horizontal axis and the maximum drain current on the vertical axis. A proper drain current must be chosen. Then draw a line from the vertical axis at that point to the load line. The point at which this current line intersects the load line is directly over the proper gate-source voltage (V_{GS}). The slope of the line from the origin (0,0) to the

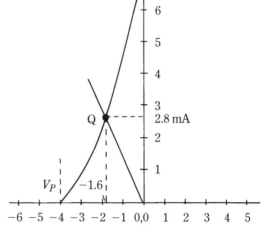

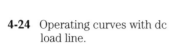

4-24 Operating curves with dc load line.

Q point represents the resistance of the correct source resistor. In this case, the slope is:

$$Q_S = \frac{0.0028 \text{ A}}{1.6 \text{ V}} \tag{4-30}$$

$$= 0.00175 \text{ S}$$

$$R_S = \frac{1}{g_m} \tag{4-31}$$

$$= \frac{1}{0.00175 \text{ S}}$$

$$= 570 \text{ }\Omega$$

Experiment 4-1

1. Connect the circuit in Fig. 4-EXP1.
2. Set an audio signal generator for a frequency of 400 Hz and set the output signal level to 0. Use the sine waveform.
3. Adjust R1 to maximum resistance (500 Ω).
4. Connect the output of this circuit (V_o) to the vertical input of an oscilloscope.
5. Advance the level control on the signal generator until a signal waveform is seen on the scope (adjust the scope controls for a full presentation).
6. Examine the waveform shown on the oscilloscope while adjusting R1. Note the effect of the setting of R1 on the amplitude and waveshape.
7. Measure the dc voltages at the source, drain, and gate of Q1.
8. Reverse the positions of the 270-Ω resistor and 500-Ω potentiometer, and then repeat this experiment.

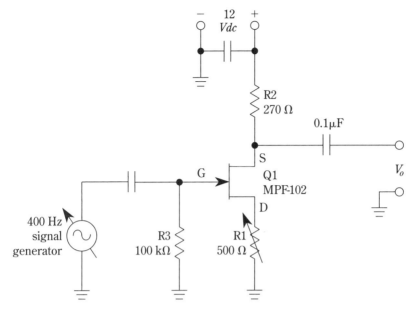

4-EXP1 Circuit for experiment 4-1.

5
Active integrated circuit devices

ONE OF THE GREAT ADVANCES IN ELECTRONICS TECHNOLOGY WAS THE INVENTION of the integrated circuit (IC) or *chip*. These devices incorporate a large number of resistors, diodes, and transistors into a single, small package, and on a common substrate. They are treated as a single entity in electronic circuits. Although there are many forms of IC device, a large portion of this book is devoted to the *operational amplifier*.

The "op amp" is very popular. It is used in a wide variety of applications, and is, perhaps, the easiest IC device that is both flexible and sufficiently well-behaved for the newcomer to use. Furthermore, the device really defines the field of linear integrated circuits. Even many supposed "non-op amps" are actually special-purpose op amps in disguise (e.g., internal connections or components that make the device work in a special way). This book concentrates on the op amp because it is ubiquitous among linear circuits, and forms the basis for many oscillator and waveform generator circuits.

Common integrated circuit package styles

The integrated circuit is formed on a tiny "chip" of silicon material by a photolithographic process. Typical chip "die" sizes range around 100 mils (0.100 inch), with some being larger and others being smaller. The die is typically mounted inside of a package and connected to the package pins by fine wires.

Figure 5-1A shows a die with wire attached, while Fig. 5-1B shows a packaged die with a see-through window for illustration purposes. The connecting wires between package pins and "solder" pads on the die are around 10 mils (0.010 inch) diameter, and are made of either gold or aluminum in most cases. Either an electric current or a thermosonic process is used to melt the end of the wire onto (and bond it with) the connecting pad on the semiconductor die.

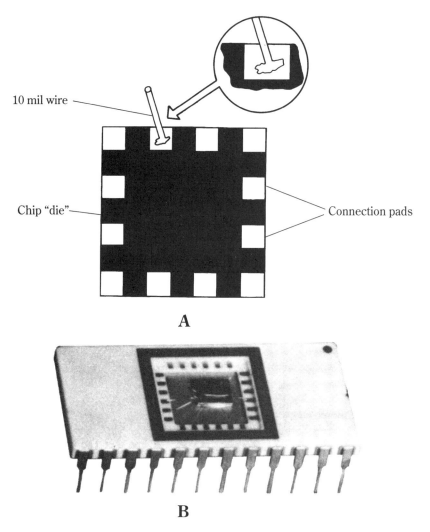

10 mil wire

Chip "die"

Connection pads

A

B

5-1 A. Attachment of leads to chip die. B. IC with die exposed.

The particular package style selected for any given IC device depends in part on the intended application and the number of pins required. For many IC devices several different packages are available, with the type-number suffix denoting the package style. The earliest IC packages were the 6-, 8-, 10-, and 110-lead metal can devices (Fig. 5-2A). These packages were redesigns of (and similar to) the TO-5 metal transistor package.

When viewed from the bottom of the package, the keyway marks the highest number lead or "pin" (e.g., pin no. 8 in Fig. 5-2A), and pin no. 1 is the next pin clockwise from the keyway. You must be careful when looking at IC base diagrams to know whether a top or bottom view is depicted.

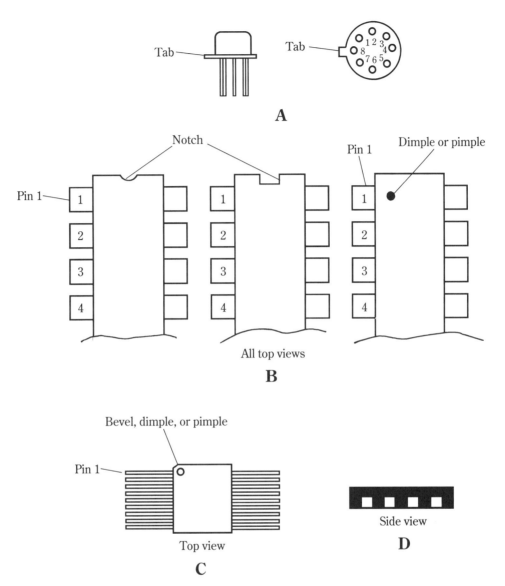

5-2 Integrated circuit packages: A. Metal package. B. Dual in-line packages (DIP). C. Flat-pack, and D. Surface mount.

Perhaps the largest number of IC devices on the market today are sold in *dual in-line packages* (DIP), examples of which are shown in Fig. 5-2B. DIP packs are available in a wide variety of sizes from 4 to more than 48 pins. Although many devices are available in other size DIP packs, most linear devices are found in 8-, 14-, or 16-pin DIP packs.

The DIP pack is symmetrical with respect to pin count regardless of the package size, so some other means is needed to designate pin no. 1. Figure 5-2B shows several common methods. In all cases the IC DIP pack is viewed from the top. In some devices a paint dot, "pimple," or "dimple" will mark pin no. 1. In other cases, a semi-circle or square notch marks the end where pin no. 1 is located. When viewed from the top, with the notch pointed away from you, pin no. 1 is to the left of the notch, while the highest numbered pin is to the right of the notch.

Both plastic and ceramic materials are used in DIP pack construction. In general, the plastic packages are used in consumer and noncrucial commercial (or industrial) equipment, while the ceramic packs are used in military and crucial commercial equipment. The principal difference between the plastic and ceramic packages is the intended temperature range. Although exceptions exist, typical temperature range-specifications are 0° to +70°C for commercial plastic devices, and −10° to + 80°C for ceramic. Military and some crucial commercial ceramic devices are rated from −55° to + 125°C. The principal difference between military chips and the highest grade commercial or industrial chips is the amount of testing, "burn-in," and documentation that accompanies each device.

An example of an IC flat-pack is shown in Fig. 5-2C. This type of package is typically used where very high component density is required. Most flat-pack devices are digital ICs, although a few linear devices are also offered. DIP package IC devices are mounted either in sockets or by insertion of the pins through holes drilled in a printed circuit board (PCB). Flat-packs, on the other hand, are mounted on the surface by direct soldering to the conductive track of the PCB.

A relatively new style of package is the *surface-mounted device* (SMD), an example of which is shown in Fig. 5-2D. SMD technology represents a significant improvement in packaging density. SMD components can be mounted closer together than other types of package, and are amenable to automatic PCB production methods. It is expected that the SMD package eventually will overtake and replace other forms, especially in VLSI applications.

Scales of integration

Several different scales of integration are found among IC devices. The ordinary *small-scale integration* (SSI) device consists of single gates, small amplifiers, and other smaller circuits. The number of components on each chip is on the order of 20 of less. *Medium-scale integration* (MSI) devices have a slightly higher degree of complexity, and may have about 100 or so components on the chip. Devices such as operational amplifiers, shift registers, counters, and so forth are usually classed as MSI devices. *Large-scale integration* (LSI) devices are mostly digital ICs and include functions such as calculators and microprocessors. Typical LSI devices contain from about 100 to 1000 components. Some newer devices are called *very large scale integration* (VLSI) devices and include some of the latest computer chips. The numbers and descriptions listed for SSI, MSI, and LSI devices are approximate only, but serve to provide guidelines. Most linear IC devices are either SSI or MSI, with the latter predominating.

Skills experiment

Integrated-circuit electronics requires the ability to solder small pins that are close together. This experiment is really just a skills exercise, but it is very important to learn.

1. Buy a grab bag of miscellaneous integrated circuits, and a perforated circuit board. The board should be designed especially for DIP IC devices. Radio Shack® has a number of these for sale, as well as souvenir bags of (probably either useless or unidentifiable) "miscellaneous" ICs (sold "as is").
2. Install a DIP IC on the board, and practice soldering it in place. Use a 25 to 75-W "pencil" soldering iron.
3. Examine your work with a magnifying glass or eyeglass. The joints should be bright, shiny, and clear of flux, and adjacent connections should not be shorted together.
4. Solder a dozen or more chips into place and repeat the inspection. Do it until you get it right.

Alternatively, obtain a large number of 8-, 14-, and 16-pin IC printed circuit sockets and do the same exercise.

Operational amplifiers

In this section, the basics of operational amplifiers are discussed. The role of these devices in analog electronics is so great that the material in this section is very important to any discussion of amplifiers.

One of the early texts on operational amplifiers tells us that the IC operational amplifier has made ". . . the contriving of contrivances a game for all." A physics professor at my college was struggling with the design of a transistor amplifier for his undergraduate students to use in a classroom experiment. It seemed very difficult to build a circuit that both had a gain of precisely 100 without the need for adjustment, and was replicable for the 30 or so amplifiers required for the class (using "on-hand" components of mixed and doubtful ancestry). After he learned about the 741 IC operational amplifier, his problems were all but over!

The little 741, although today not even considered for many jobs because of the newer types now on the market, then cost just over a dollar, yet it obeyed very simple design equations. We can design op-amp voltage amplifiers by using only the ratio of two resistors forming a voltage-divider feedback network.

The operational amplifier originally was designed to perform mathematical operations in analog computers (hence the name "operational" amplifier). The first op amps were vacuum-tube models and only approximated the behavior of the ideal mathematical model of the device. Later, transistor operational amplifiers were available, and finally integrated-circuit types came on the market. One of the first linear ICs made commercially was the μA-709 operational amplifier. Although now considered primitive, that device sold for $109 in one mid-sixties catalog (current price is about five for a dollar—where they are still available).

Figure 5-3A shows the usual circuit symbol of the operational amplifier. An alternate symbol shown in Fig. 5-3B is preferred by some people. This symbol is technically the correct symbol to use. That version uses a curved back to which the input leads are attached. The version shown in Fig. 5-3A is used almost universally, however, even though it technically denotes any amplifier—including, but not limited to, operational amplifiers. We will maintain the de facto standard of Fig. 5-3A.

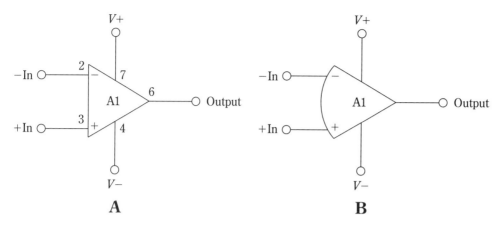

5-3 Operational amplifier symbols: A. Common. B. Technically correct but rarely used.

Note the pin-outs for the amplifiers in Fig. 5-3. The pin numbers given are for the 741 device, but they have become something of an industry standard. There are two input connections, two power-supply connections, and one output connection. There is no ground or common connection. The ground and signal common are taken from the power-supply common line.

The two power-supply connections are $V+$ and $V-$. The $V+$ supply is positive with respect to common, while the $V-$ is negative with respect to common. The range for these voltages is typically ± 4 to ± 18 V, although a number of examples exist with wider (or slightly different) voltage ranges. An RCA CA-3140 BiMOS device, for example, operates at potentials up to ± 22 V for $V-$ and $V+$, while certain "low-power" operational amplifiers operate down to ± 1.5 Vdc.

In addition to the absolute voltage limits, there are also sometimes relative limitations. For example, other 741 devices had a 30-V limit for the voltage defined by the expression $[(V+) - (V-)]$, even though each $V-$ and $V+$ can be as high as 18 V. Thus, if $V+$ is $+18$ V, then $V-$ must be not greater than -12 V in order that the differential not be greater than 30 V $[(+18)-(-12) = +30 \text{ V}]$.

The selection of power-supply voltages might also depend somewhat on the maximum anticipated output voltage. If the amplifier is being designed for use with an analog-to-digital converter that has a range of -10 to $+10$ V input, then we certainly want the output of the amplifier to be capable of achieving those limits.

However, there is a limit on how high the output voltage can reach, and that limit is a function of the power-supply voltage. In general, the limitation is based on the number of pn junctions between the output terminal on the IC and the two power-supply terminals. Each pn junction has a 0.7-V drop that must be accounted for. If there are four pn junctions between the output terminal and the $V+$ power-supply terminal, for example, the maximum allowable output voltage will be $[(V+) - (4 \times 0.7)]$ V, or 2.8 V lower than $V+$. Thus, when we want the output terminal to swing to $+10$ V, the absolute minimum $V+$ power-supply voltage will be $10 + 2.8$ V, or $+12.8$ Vdc. Obviously, a $+12$ Vdc power supply will not work in this case.

In general, ordinary bipolar transistor operational amplifiers require a supply voltage of 2 to 4.5 V higher than the maximum required output voltage, yet must remain within the $V+$ and $V-$ constraints of the device. Some BiMOS and BiFET devices are available in which the maximum output signal voltage can be as close as 0.5 V to the dc power-supply potential.

The two inputs for the operational amplifier form a so-called *differential pair* because they operate 180° out of phase with each other. The inverting input produces a 180° phase shift between the input signal and the output signal. (In other words, a positive-going input signal produces a negative-going output signal, and vice versa.) The noninverting input produces a 0° phase shift in the output signal. Since one input produces an in-phase output and the other produces an out-of-phase output, application of the same voltage to both inputs simultaneously produces a zero net output potential. We will use this fact in a later section to form the immensely useful differential amplifier.

The two inputs on the operational amplifier have a very high impedance (which is infinite in the ideal model). Thus, they are a perfect voltage-amplifier input.

The output of the operational amplifier is also suited to use as a voltage-amplifier circuit. The output impedance of the typical op amp is usually quite low (50 to 200 Ω), so forms a nearly perfect voltage source.

Operational-amplifier dc power supplies

Figure 5-4 shows a model of the typical operational-amplifier power supply. Although batteries are shown here for sake of simplicity, electronic power supplies operated from the ac power lines are commonly used. Recall that we have two different

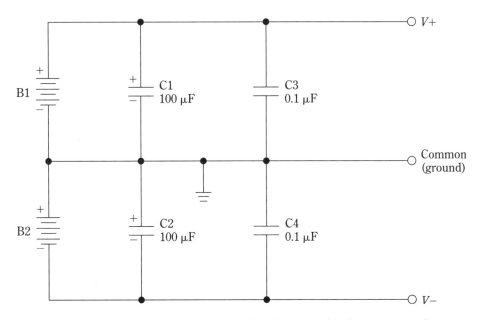

5-4 Operational-amplifier dc power supply scheme uses bipolar power supplies.

voltages in the op-amp power supply: *V*+ and *V*−. Voltage *V*+ is supplied by battery B1, while *V*− is supplied by battery B2. The common (or ground) connection is the junction between the two batteries. Normally, B1 and B2 will have the same voltage rating, although that is not strictly a requirement unless other circuit considerations apply.

The capacitors shown in Fig. 5-4 are used for decoupling, especially when multiple stages are fed from the same power supply. Capacitors C1 and C2 are normally 50 to 200 μF, and are used for decoupling low-frequency signals. Capacitors C3 and C4 are generally 0.1 μF and are used for decoupling higher frequency signals. We cannot normally use the higher value C1 and C2 for high-frequency signals because these are normally electrolytic capacitors, which are ineffective at high frequencies.

The signal common or "ground" connection is used as the zero reference for input and output signals on the operational amplifier. Whether it is actually grounded or not depends upon circuit design considerations. In most cases, however, it is grounded for the sake of simplicity.

In most applications, electronic power supplies used for B1 and B2 must be voltage regulated. Although there are certainly numerous applications where voltage-regulated dc power supplies are not strictly required, they are almost always "good engineering practice."

The ideal operational amplifier

Before getting further into operational-amplifier circuits, let's set the stage for our simplistic circuit analysis by discussing the properties of the so-called "ideal operational amplifier." This ideal device has the following properties:

1. Infinite open-loop gain.
2. Zero output impedance.
3. Infinite input impedance.
4. Zero noise contribution.
5. Infinite bandwidth.
6. Differential inputs "stick together."

What do these statements mean, and how do they compare with real, practical IC operational amplifiers?

Infinite open-loop gain   *Infinite open-loop gain* means that the voltage gain of the ideal operational amplifier in the open-loop (i.e., no feedback) configuration is infinite. Real operational amplifiers do not even approach the ideal, but are still good enough approximations to make the device behave properly. The ability of the practical operational amplifiers to approach the ideal is dependent upon having an extremely high open-loop gain (otherwise, the equations behave badly). In practical devices, the open-loop voltage gain—A_{vol}—is 20,000 in low-cost devices, and well over 1,000,000 in premium devices.

Zero output impedance   The operational amplifier is supposed to be a perfect voltage amplifier, so requires a zero output impedance. Real devices have output impedances of 50 to 200 Ω, with most being under 100 Ω.

Infinite input impedance   *Infinite input impedance* means that the input will neither sink nor source electrical current. Recall that input impedance is $Z_{in} = V_{in}/I_{in}$, so for input impedance to be infinite, I_{in} must be 0. In real operational amplifiers, of course, we find this value is nonzero and is one of the primary differences between premium and low-cost devices. Low-cost amplifiers have input (transistor) bias currents to contend with of up to 1 or 2 mA, but certain other devices measure the input current in picoamperes or nanoamperes. The RCA BiMOS operational amplifiers (CA-3140, etc.) use MOSFET input transistors to produce an input impedance of more than 10^{12} Ω.

Zero noise contribution   The *noise* referred to is internal noise added to the signal. This is one area of primary difference between the low-cost and premium devices. The low-cost amplifiers add considerable "hiss" noise, so are unusable on low-signal applications.

Infinite bandwidth   *Infinite bandwidth* means that there is no limit to the operating frequency of the device, which in real operational amplifiers is patently absurd. Unconditionally stable, frequency-compensated devices like the 741 may have an upper frequency limit of only a few kilohertz, while other operational amplifiers operate to several megahertz. Only a few high-HF or low-VHF devices are available, and they are usually labeled "video operational amplifiers," or something similar.

Differential inputs stick together   This property is essential to the simplified circuit analysis which we use. The implication of this property is that a voltage applied to one input will also appear on the other input. We must treat both inputs alike mathematically in this regard. Thus, if we apply a voltage to the noninverting input, we must treat the inverting input as if it also sees that voltage. This is not just some theoretical device used to make equations work. If you apply a real voltage to a real noninverting input, and then use a real voltmeter at the inverting input, you will measure a real voltage at that point.

We will use these properties in the circuit descriptions that follow.

Standard operational-amplifier parameters

Designing circuits with operational amplifiers requires understanding the various parameters given in the specification sheets. The list below is not intended to be exhaustive, but rather it represents the most commonly needed parameters.

Open-loop voltage gain (A_{vol})   *Voltage gain* is defined as the ratio of output signal voltage to input signal voltage (V_o/V_{in}), which is a dimensionless quantity. The open-loop voltage gain is the gain of the circuit without feedback (i.e., with the op amp's external feedback loop open-circuited). In an ideal operational amplifier, A_{vol} is infinite, but in practical devices it will range from about 20,000 for low-cost devices to over 1,000,000 in premium devices.

Large signal voltage gain   *Large-signal voltage gain* is defined as the ratio of the maximum allowable output voltage swing (usually several volts less than $V-$ and $V+$) to the input signal required to produce a swing of ± 10 V (or some other standard).

Slew rate   *Slew rate* specifies the ability of the amplifier to transition from one output-voltage extreme to the other extreme, while delivering full rated output current to the external load. The slew rate is measured in terms of voltage change per unit of time. The 741 operational amplifier, for example, is rated for a slew rate of 0.5 V per microsecond (0.5 V/μs). Slew rate is usually measured in the unity-gain noninverting follower configuration (of which, more later).

Common-mode rejection ratio (CMRR)   A *common-mode voltage* is one that is presented simultaneously to both inverting and noninverting inputs (voltage V3 in Fig. 5-5). In an ideal operational amplifier, the output resulting from the common-mode

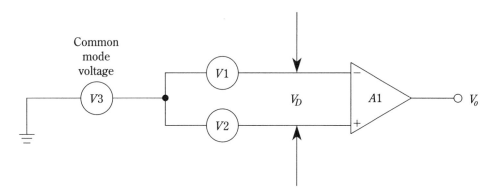

5-5 Signal voltages applied to an op amp.

voltage is zero, but in real devices it is nonzero. The *common-mode rejection ratio* (CMRR) is the measure of the device's ability to reject common-mode signals, and is expressed as the ratio of the differential gain to the common-mode gain. The CMRR is usually expressed in decibels (dB), with common devices having ratings between 60 dB and 120 dB (the higher the number, the better the device).

Power-supply rejection ratio (PSRR) Also called *power supply sensitivity*, the PSRR is a measure of the operational amplifier's insensitivity to changes in the power-supply potentials. The PSRR is defined as the change of the input offset voltage (see below) to a 1-V change in one power-supply potential, while the other is held constant. Typical values are in microvolts, or millivolts, per volt of power-supply potential change.

Input offset voltage The voltage required at the input to force the output voltage to zero when the signal voltage is zero is called the *input offset voltage*. The output voltage of an ideal operational amplifier is 0 when $V_{in} = 0$.

Input bias current *Input bias current* is the current flowing into or out of the operational-amplifier inputs. In some sources, this current is defined as the average difference between currents flowing in the inverting and noninverting inputs.

Input offset (bias) current *Input offset current* is defined as the difference between inverting and noninverting input bias current when the output voltage is held at zero.

Input signal voltage range The range of permissible input voltages as measured in the common-mode configuration—i.e., the maximum allowable value of V3 in Fig. 5-5.

Input impedance The *input impedance* is the resistance between the inverting and noninverting inputs (Z_{in} in Fig. 5-6). This value is typically very high: 1 MΩ in low-cost bipolar operational amplifiers and over 10^{12} Ω in premium BiMOS devices.

Output impedance The *output impedance* is the "resistance looking back" (to borrow a concept from Thevinin's theorem) into the amplifier's output terminal and is usually modeled as a resistance in series with the output signal and ground (despite the lack of a ground terminal on the device). Typically the output impedance is less than 100 Ω.

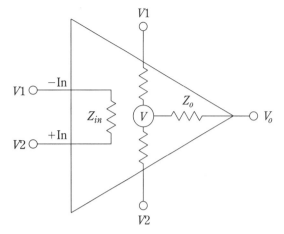

5-6 Equivalent circuit of operational amplifier.

Output short-circuit current The current that will flow in the output terminal when the output load resistance external to the amplifier is 0 Ω (i.e., a short to ground) is called the *output short-circuit current*.

Channel separation *Channel separation* is a parameter used on multiple operational-amplifier integrated circuits—i.e., two or more operational amplifiers sharing the same package with common power-supply terminals. The separation specification tells us something of the isolation between the op amps inside the same package and is measured in decibels. The 747 dual operational amplifier, for example, offers 120 dB of channel separation. From this specification we may imply that a 1 μV change will occur in the output of one of the amplifiers when the other amplifier output changes by 1 V (1 V/1 μV = 120 dB).

Minimum and maximum parameter ratings

Operational amplifiers, like all electronic components, are subject to certain maximum ratings. If these ratings are exceeded, the user can expect premature—often immediate—failure, or unpredictable operation. The following ratings are the most commonly used.

Maximum supply voltage The *maximum supply voltage* is the maximum potential that can be applied to the operational amplifier without damaging the device. The operational amplifier uses $V+$ and $V-$ dc power supplies that are typically ±15 Vdc, although some exist with much higher maximum potentials. The maximum rating for either $V-$ or $V+$ often depends upon the value of the other.

Maximum differential supply voltage The *maximum differential supply voltage* is the algebraic sum of $V-$ and $V+$, namely $[(V+) - (V-)]$. It is often the case that this rating is not the same as the summation of the maximum supply voltage ratings. For example, one 741 operational-amplifier specification sheet lists $V-$ and $V+$ at 15 V each, but the maximum differential supply voltage is only 28 V. Thus, when both $V-$ and $V+$ are at maximum (i.e., 15 Vdc each), the actual differential supply voltage is $[(+15 \text{ V}) - (-15 \text{ V})] = 30$ V, which is 2 V over the maximum rating. Therefore, when either $V-$ or $V+$ is at maximum value, the other must be proportionally lower. For example, when $V+$ is +15 V, the maximum allowable value of $V-$ is [28 V − 15 V] = 13 V.

Power dissipation The P_d rating is the *maximum power dissipation* of the operational amplifier in the normal ambient temperature range (80°C in commercial devices, and 125°C in military-grade devices). A typical rating is 500 mW (0.5 W).

Maximum power consumption The maximum power required, usually under output short-circuit conditions, that the device will survive is called the *maximum power consumption*. This rating includes both internal power dissipation and device power requirements.

Maximum input voltage The *maximum input voltage* is the maximum potential that can be applied simultaneously to both inputs. Thus, it is also the maximum common-mode voltage. In most bipolar operational amplifiers, the maximum input voltage is equal to the power-supply voltage, or nearly so. There is also a maximum input voltage that can be applied to either input when the other input is grounded.

Differential input voltage The *differential input voltage* rating is the maximum differential-mode voltage that can be applied across the inverting (−) and noninverting (+) inputs.

Maximum operating temperature The *maximum operating temperature* is the highest ambient temperature at which the device will operate according to specifications with reasonable reliability. The usual rating for commercial devices is 70° or 80°C, while military components must operate to 125°C.

Minimum operating temperature The *minimum operating temperature* is the lowest temperature at which the device operates within specifications. Commercial devices operate down to either 0° or −10°C, while military components operate down to −55°C.

Output short-circuit duration The length of time the operational amplifier will safely sustain a short circuit of the output terminal is called the *output short-circuit duration*. Many modern operational amplifiers are rated for indefinite output short-circuit duration.

Maximum output voltage The output potential of the operational amplifier is related to the dc power supply voltages. Most operational amplifiers have several bipolar pn junctions between the output terminal and either $V-$ or $V+$ terminals, and the voltage drop across these junctions reduces the *maximum output voltage*. For example, if there are three pn junctions between the output and power-supply terminals, the maximum output voltage is $[(V+) - (3 \times 0.7)]$, or $[(V+) - 2.1]$ V. If the maximum $V+$ voltage permitted is 15 V, the maximum allowable output voltage is $[(15 \text{ V}) - (2.1 \text{ V})]$, or 12.9 V. It is not always true that the maximum negative output voltage is equal to the maximum positive output voltage. A related rating is the maximum output voltage swing, which is the absolute value of the voltage swing from maximum negative to maximum positive.

Inverting follower circuits

Figure 5-7 shows the inverting follower circuit. In this circuit, the noninverting input is grounded, so we must treat the inverting input as if it were also grounded (recall ideal property number 6). This fact gives rise to a somewhat confusing concept: *virtual ground*. The inverting input is not actually grounded, but since it is at zero po-

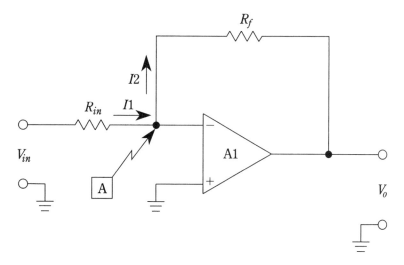

5-7 Inverting follower circuit.

tential because the other input is grounded, we say that it is "virtually grounded." The concept is simple, only the semantics are confounding.

Let's consider the currents appearing in node "A" of Fig. 5-7. We know from property number 3 that I_3, the input bias current, is zero. We also know from Kirchhoff's Current Law (KCL) that all currents into and out of a junction algebraically sum to zero. Thus,

$$I_1 = -I_2 \qquad (5\text{-}1)$$

We also know from Ohm's law that:

$$I_1 = \frac{V_{in}}{V_o} \qquad (5\text{-}2)$$

and,

$$I_2 = \frac{V_o}{R_{in}} \qquad (5\text{-}3)$$

Thus, when we substitute these two equations into KCL above:

$$\frac{V_{in}}{R_{in}} = \frac{-V_o}{R_f} \qquad (5\text{-}4)$$

We know that a voltage amplifier's transfer function is V_o/V_{in}, so solving the above equation for the transfer function yields:

$$\frac{V_o}{V_{in}} = \frac{-R_f}{R_{in}} \qquad (5\text{-}5)$$

Thus, the voltage gain, A_v of the inverting follower is given by the ratio of two resistors:

$$A_v = \frac{-R_f}{R_{in}} \qquad (5\text{-}6)$$

We may, then, design the inverting amplifier simply by manipulating the values of these two resistors (see Fig. 5-7).

There is sometimes a constraint on the minimum allowable value of R_{in}. Point "A" is essentially grounded, so the input impedance of this circuit is simply the resistance of R_{in}. There is a rule of thumb in voltage-amplifier design that says the input impedance of a stage must be five times (most prefer ten times) the source impedance of the signal source.

Experiment 5-1

Design and build a gain-of-100 inverting amplifier that has an input impedance of 10 kΩ or more. Because the input impedance must be 10 kΩ, R_{in} must be 10 kΩ or more. Let's set it at 10 kΩ. Solving the gain equation for R_f yields:

$$R_f = A_v R_{in}$$
$$= (100)\,(10{,}000\ \Omega)$$
$$= 1{,}000{,}000\ \Omega$$

Thus, a 10-kΩ input resistor and a 1-MΩ feedback resistor yields a gain of 100. Since this is an inverting follower, the gain is actually −100. (NOTE: The "−" sign indicates that a 180° phase reversal of the signal takes place between input and output.)

Now that you've designed the circuit, build it and test it to compare the results with the theory.

Would you like to test the "virtual ground" idea? Set the signal generator output to 0. Use a dc voltmeter to measure the potential at the inverting and noninverting input pins of the op-amp IC.

Noninverting followers

The noninverting follower applies a signal to the noninverting input. There are two basic configurations for the noninverting follower:

- Unity gain noninverting follower.
- Noninverting follower with gain.

Figure 5-8A shows the unity-gain noninverting follower. The output terminal is connected to the inverting input, producing 100 percent feedback. The output voltage is equal to the input voltage. So what use is a unity-gain (i.e., gain of 1) voltage amplifier? There are three uses: *buffering, impedance transformation,* and *power amplification.*

Buffering means using an amplifier to isolate a circuit from its load. Some transducers and most oscillators or astable multivibrators will change frequency if the load impedance changes, so a unity-gain noninverting follower will help "buffer" the circuit.

Impedance transformation occurs because the input impedance is terribly high, while the output impedance is very low. We can use this circuit, for example, when acquiring a signal from biological or chemical sensor sources where the natural impedance is extremely high. A pH electrode, for example, has source impedances ranging from 10 MΩ to 100 MΩ.

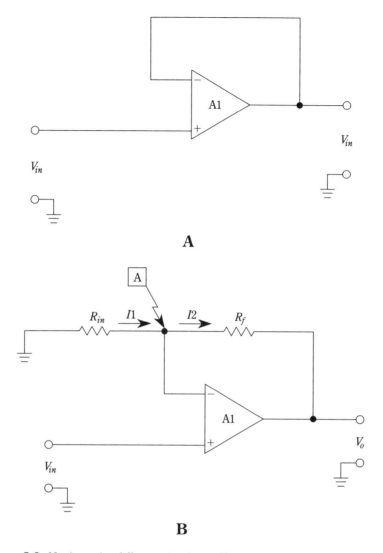

5-8 Noninverting follower circuits: A. Unity gain. B. With >1 gain.

The power amplification is small, but is illustrated by the fact that the voltage remains constant ($V_o = V_{in}$), but the impedances are unequal. Obviously, since power is defined by $P = V^2/R$, reducing R while keeping V the same results in higher power (P).

Experiment 5-2

1. Build the circuit of Fig. 5-EXP2.
2. Using either an oscilloscope or ac voltmeter, measure the input and output signal voltages.
3. Compare the input and output voltages by taking the ratio V_o/V_{in}.

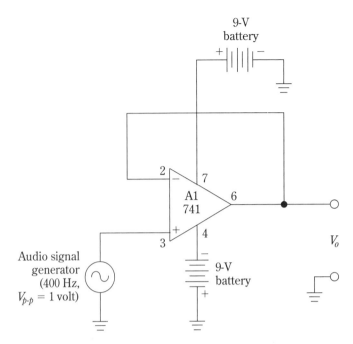

5-EXP2   Circuit for Experiment 5-2.

The noninverting follower with gain circuit of Fig. 5-8B retains the properties of the unity-gain circuit, but produces voltage gain as well. Keeping in mind that the inverting input (point "A") is at V_{in}, analysis similar to the previously used method produces a voltage gain of:

$$A_V = \frac{R_f}{R_{in}} + 1 \qquad \qquad (5\text{-}7)$$

The noninverting follower circuits are used wherever extremely high input impedance is needed or where no phase reversal can be tolerated.

Experiment 5-3

1. Build the circuit of Fig. 5-EXP3.
2. Using either an oscilloscope or ac voltmeter, measure the input and output signals voltages.
3. Compare the input and output voltages by taking the ratio V_o/V_{in}.
4. Set the signal-generator level to 0, and then use a dc voltmeter to measure the voltages appearing at the inverting and noninverting input pins of the op-amp IC.

Operation from a single dc power supply

Operational amplifiers are normally powered from a bipolar dc power supply. Such a power supply has $V+$ and $V-$ voltages that are each referenced to ground or common. This system essentially requires two semi-independent dc supplies. In some

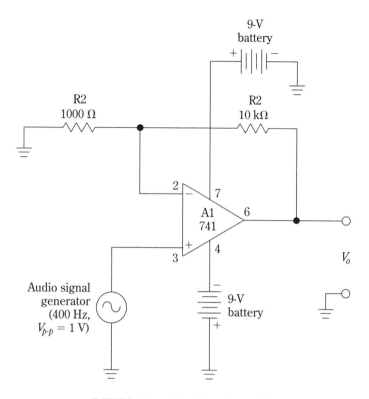

5-EXP3 Circuit for Experiment 5-3.

cases, either ultimate use or other design constraints force the use of a single, monopolar dc power supply. In this section we will discuss simple methods for operating the amplifier from a single dc power supply.

Some schemes exist for creating a split power supply from a monopolar supply in order to mimic bipolar power-supply operation. One such scheme connects two zener diodes in series across the single supply, along with the necessary current-limiting resistors. The junction between the two zener diodes becomes the signal common. A limitation of this method is that the dc supply cannot be chassis referenced.

Another scheme is to use the regular monopolar dc power supply for $V+$ and then use a dc-to-dc converter circuit for $V-$. Such a circuit is little more than an ac oscillator in the 20 to 500 kHz range, with its output signal rectified and filtered to produce the $V-$ voltage.

Figure 5-9A shows the method for biasing the operational amplifier inputs to permit single-supply operation. This technique is based on the simple resistor voltage-divider circuit of Fig. 5-9A. The output voltage (V_1) is given by the standard voltage divider equation:

$$V_1 = \frac{R_2 \, (V+)}{R_1 + R_2} \qquad \textbf{(5-8)}$$

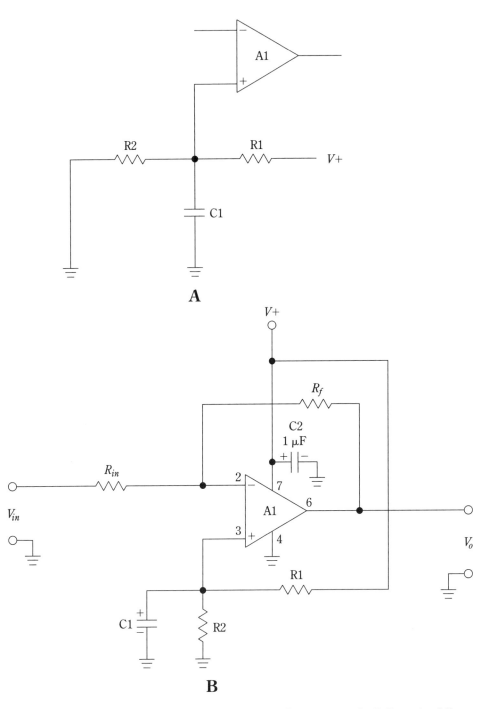

5-9 A. Bias network for operating op amp from single dc power supply. B. Inverting follower circuit using one dc power supply.

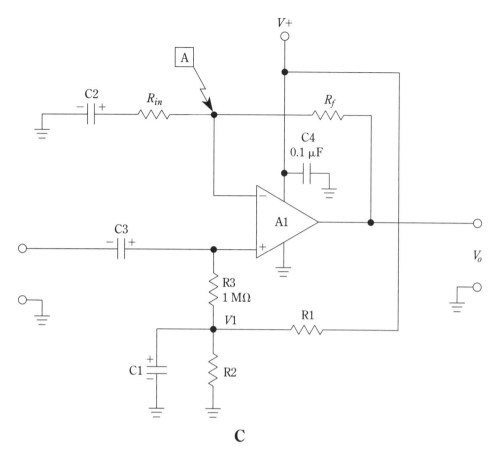

5-9C Noninverting follower using one dc power supply.

In most cases, the value of V_1 will be one-half V+, so that the operational amplifier has a quiescent output point that is midway between extremes. This bias level is achieved by making $R_1 = R_2$. The value of R_1 and R_2 is usually selected such that it falls between 1 kΩ and 100 kΩ.

The capacitor shunting resistor R2 (i.e., C1) decouples ac variations. The value of capacitance is selected for a reactance value of one-tenth R_2 (i.e., $R_2/10$) at the lowest frequency of operation. For example, suppose $R_2 = 10$ kΩ, and the lowest frequency of operation is 10 Hz. If R_2 is 10 kΩ, then the capacitive reactance of the shunt capacitor should be: $R_2/10 = (10$ kΩ$)/10 = 1$ kΩ. Solving the usual capacitive reactance equation for C_1 gives us:

$$C_1 = \frac{10^6}{2\,\pi\,FX_c} \qquad\qquad (5\text{-}9)$$

$$= \frac{1{,}000{,}000}{(2)\,(3.14)\,(10\text{ Hz})\,(1000\ \Omega)}$$

$$= \frac{1,000,000}{62,800}$$
$$= 15.9 \ \mu F$$

The value 15.9 μF is nonstandard, so a 20- or 22-μF unit would normally be selected as a practical matter.

Figure 5-9B shows the method for biasing an operational amplifier in the inverting-follower configuration. In the bipolar-supply version of this circuit (Fig. 5-7), the noninverting input is grounded (i.e., set to 0 V). In single-supply operation, however, we apply bias voltage V_1 to the noninverting input. This voltage (V_1) will also appear on the inverting input (hence the need for dc blocking capacitor C2). The output terminal will be biased up according to the value of V_1, so might require a dc blocking capacitor also (shown in Fig. 5-9C) if such a voltage adversely affects the following stage.

The value of capacitor C2 is selected to have a low impedance at the lowest frequency of operation, using a protocol similar to that discussed above for the voltage-divider shunt capacitor. A general rule of thumb is to regard $R_{in}C_1$ as a high-pass filter with a cut-off frequency, F_c, equal to $1/(2\pi R_{in}C_2)$. The object is to select a value of C_2, given a value for R_{in}, that results in a value of F_c lower than the lowest operating frequency.

The circuit configuration for noninverting follower circuits is shown in Fig. 5-9C. This circuit is the same as for inverting followers except for resistor R3. The purpose of R3 is to maintain a high input impedance to signals applied to the noninverting input. The minimum value of $R3$ is at least ten times the output resistance of the driving stage. In practical cases, however, the source impedance is usually low enough that it is possible to set R_3 to 100 or 1000 times the source impedance. Typical values range from 10 kΩ to 1 MΩ, with 100 kΩ predominating.

The value of C_1 is selected such that the cut-off frequency of the filter formed by C_1R_3 is lower than the lowest operating frequency. The same equation applies here as previously.

Capacitor C4 and resistor R4 are used when the $(V+)/2$ bias on the output terminal will adversely affect a following stage or instrument. Again, the "lowest frequency of operation" rule is invoked when setting the value of C_1, with the resistance being the input resistance of the stage following.

Experiment 5-4

1. Build the circuit of Fig. 5-9A, using a 741 operational amplifier (pinouts shown in preceding experiments). Use a value of resistance between 1 kΩ and 20 kΩ for the voltage-divider resistors. Use 9 Vdc batteries or ±12 Vdc electronic power supplies.
2. Measure the dc voltages on the noninverting input, the inverting input, and the output pins of the operational amplifier.
3. Replace the resistor voltage divider with a potentiometer. Connect one end of the potentiometer to ground, one end to V+, and the wiper to the resistor leading to the noninverting input of the op amp.
4. Repeat the preceding measurements.

Op-amp problems and their solution

Earlier we discussed the concept of the ideal operational amplifier. Such a hypothetical device is merely a learning tool, and doesn't really exist. While it makes our analysis easier, it cannot be purchased and used in practical circuits. All real operational amplifiers depart somewhat from the ideal, and the quality of the device is sometimes measured by the degree of that departure. We find, for example, that open-loop gain is not really infinite, but rather it is a value from 20,000 to over 1,000,000. Similarly, real operational amplifiers don't have infinite bandwidth, and in fact some are heavily bandwidth limited. The latter statement is especially true of "unconditionally stable" or "frequency compensated" operational amplifiers such as the 741 device.

While the stability is highly desirable, it is obtained at the expense of frequency response. In this section we will deal with some of the more common problems in real devices, and their solutions.

Offset cancellation

The ideal operational amplifier will produce a zero output voltage when the two inputs are at the same potential. However, real operational amplifiers often produce an offset potential at the output. These voltages are those which exist on the output terminal when it should be zero.

There are several sources. One is the input bias currents used to operate the transistors in the input stage of the operational amplifier. Typically, these currents are nonzero. When the current flows into or out of the inverting input, it will create a voltage drop equal to the product of the current and the parallel combination of R_{in} and R_f. How do we deal with this problem? The voltage drop produced by the input bias currents is amplified and appears at the output as an offset potential.

Figure 5-10 shows the use of a compensation resistor, R_c. This resistor has a value equal to the parallel combination of the other two resistors. Since the same bias current flows in both inputs, this resistor will produce the same voltage drop at the noninverting input as appears at the inverting input. Since the inputs are differential, the net output voltage is zero.

Figure 5-11 shows two methods for nulling output offsets, regardless of the source. Figure 5-11A shows the use of offset null terminals that are found on some operational amplifiers. A potentiometer is placed between the terminals, while the wiper is connected to the $V-$ power supply. This potentiometer is adjusted to produce the null required. The input terminals are shorted together, and the pot is adjusted to produce 0 V output.

Figure 5-11B shows a circuit that can be used on any operational amplifier, inverting or noninverting, except the unity-gain noninverting follower. A counter-current (I3) is injected into the summing junction (point "A") of a magnitude and polarity to cancel the output offset voltage. The voltage at point "B" is set to produce a null offset at the output. The output voltage component due to this voltage is:

$$V_{o'} = \frac{-V_B R_f}{R_2} \qquad\qquad \textbf{(5-10)}$$

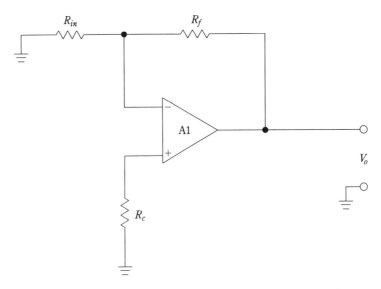

5-10 Use of compensation resistor for input offset current nulling.

If greater (i.e., finer) control over the output offset is required, then one of the two circuits of Fig. 5-12 can be used to replace the potentiometer of Fig. 5-11B.

Power-supply decoupling

Operational amplifiers, like all other active electronic components, are somewhat affected by variations in power-supply voltage and noise signals riding on the power-supply potentials. It is also possible that a signal from one operational amplifier is coupled to others over the power-supply lines.

The cure for these problems, and certain stability problems which we will discuss in due course, is the decoupling scheme shown in Fig. 5-13. The V– and V+ power supply lines are each decoupled with two capacitors. C1 and C2 are 0.1 μF capacitors and are used for high-frequency signals; C3 and C4, on the other hand, are higher valued and are for low-frequency signals. Two capacitors are used because electrolytic capacitors used for C3 and C4 are ineffective at high frequencies.

The decoupling capacitors should be mounted as close as possible to the body of the operational amplifier. This constraint is especially true when the operational amplifier is not frequency compensated and has a high gain-bandwidth product.

Frequency stability

Operational amplifiers that are not internally frequency compensated are susceptible to oscillations. Figure 5-14 shows a plot of the open-loop phase shift versus frequency for a typical operational amplifier. From dc to a certain frequency there is essentially zero phase-shift error, but above that breakpoint the phase error increases rapidly. This change is due to the internal resistances and capacitances of the ampli-

5-11 Using potentiometer and op amps
nulling terminals to cancel output offset
voltage: A. Use of offset null terminals.
B. Universal nulling circuit.

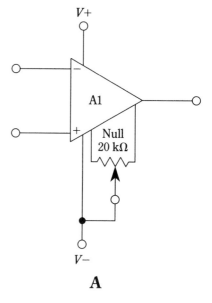

A

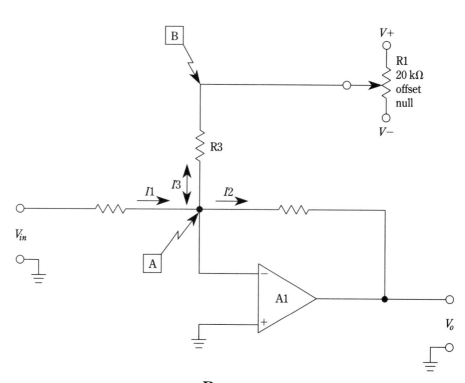

B

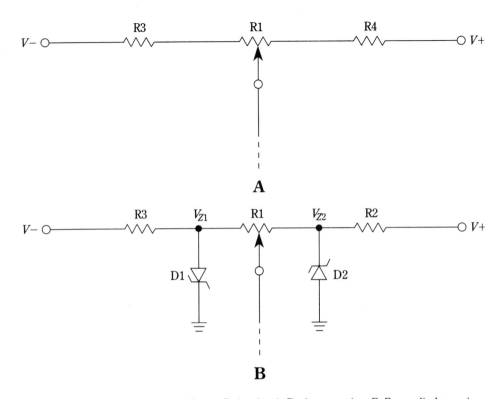

5-12 Alternative fine resolution offset null circuits: A. Resistor version. B. Zener diode version.

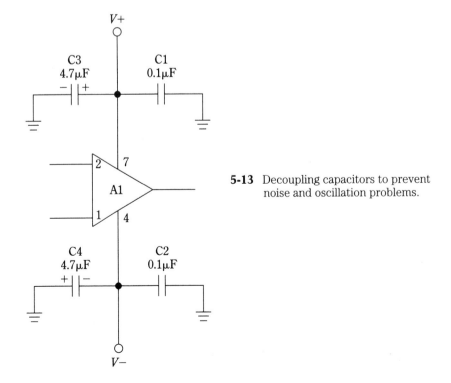

5-13 Decoupling capacitors to prevent noise and oscillation problems.

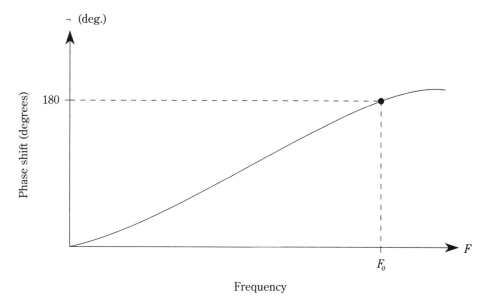

5-14 Phase shift vs. frequency curve.

fier acting as a phase-shift network. At some frequency the phase-shift error reaches 180°, which when added to the 180° inversion normal to inverting follower amplifiers adds up to the 360° phase shift that satisfies Barkhausen's criteria for oscillation. At this frequency the amplifier becomes an oscillator.

The use of power-supply decoupling helps somewhat for this problem, and it is considered poor engineering practice to use an uncompensated operational amplifier without those decoupling capacitors. In other cases, you might need to use a variant of the methods shown in Fig. 5-15.

In Fig. 5-15A we see lead compensation. If the operational amplifier is equipped with compensation terminals (usually pins 1 and 8 on "standard" packages), then connect a small-value capacitor (20 to 100 pF) as shown. An alternate scheme is to connect the capacitor from a compensation terminal to the output terminal.

The recommended capacitance in manufacturers' specification sheets is for the unity-gain noninverting follower configuration. For a gain follower the capacitance is reduced by the feedback factor, β:

$$C = C_m\beta \qquad\qquad (5\text{-}11)$$

Where: C is the required capacitance
$\quad\quad C_m$ is the recommended unity-gain capacitance
$\quad\quad \beta$ is the feedback factor $R_{in}/(R_{in} + R_f)$

Lag compensation is shown in Fig. 5-15B. In this case we connect either a single capacitor or a resistor-capacitor network from the compensation terminal to ground. A related method places the resistor-capacitor series network between the inverting and noninverting input terminals.

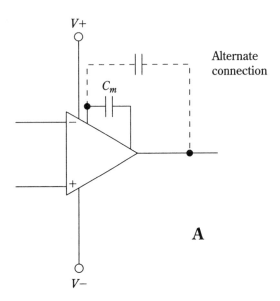

A

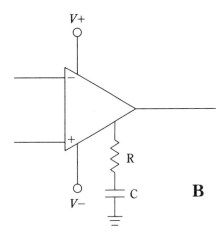

B

5-15 Frequency compensation methods: A. Lead B. Lag.

The object of these methods (see Fig. 5-16) is to reduce the high-frequency loop gain of the circuit to a point where the total loop gain is less than unity at the frequency where the 180° phase shift occurs (F_o). The amount of compensation required to accomplish this goal determines the maximum amount of feedback that can be used without violating the stability requirement.

dc differential amplifiers

A *differential amplifier* is one that produces an output voltage that is the product of the gain and the difference between the two input voltages. Figure 5-17 shows a

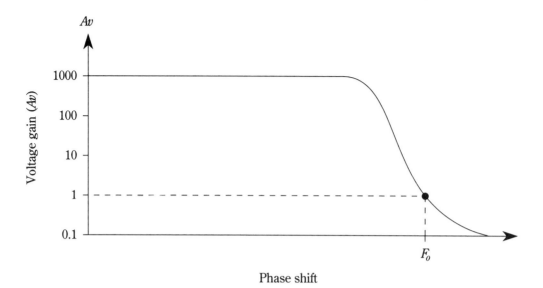

5-16 Voltage gain vs. phase shift curve.

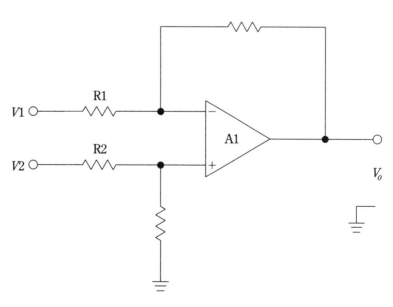

5-17 dc differential amplifier circuit.

simple dc differential amplifier that is based on a single operational amplifier. The gain is:

$$A_{vd} = \frac{R_3}{R_1} = \frac{R_4}{R_2} \qquad (5\text{-}12)$$

Provided that $R_1 = R_2$ and $R_3 = R_4$.

The output voltage from the dc differential amplifier is the product of the differential voltage gain and the difference between the two input voltages $(V_2 - V_1)$. This difference potential is known as a *differential mode signal*, and the differential amplifier ideally responds only to such signals. A common-mode signal is one that is applied to both inputs simultaneously. In the ideal differential amplifier, the common-mode gain is zero, so the common-mode signal causes no change in the output signal. Instrumentation applications, including many in the data-acquisition field, are ideal for differential amplifiers. Consider a case where a low-level signal must pass over wires in an environment that is intense with 60-Hz ac fields from the power lines. In that case, you would normally expect the signal to be obscured with 60-Hz noise. If you use a differential amplifier, however, you can make the design such that the signal voltage is a differential mode potential. The lines from the signal source can be made equal length and pass through the same environment. In that case, both lines would receive the same 60-Hz signal levels, in phase, and that noise signal is therefore common mode. The differential amplifier will reject the interfering signal.

Good common-mode rejection is possible in the circuit of Fig. 5-17, but only if the resistors are matched ($R_1 = R_2$ and $R_3 = R_4$ to a close tolerance). An improved circuit is shown in Fig. 5-18. Here, we replace resistor R4 with a series combination of a fixed resistor and a potentiometer (or just a potentiometer, in some cases). Potentiometer R5 becomes a common-mode adjust control. It is adjusted with $V_1 = V_2$ (which means shorted together and applied to a single signal source) for minimum output voltage.

6
Relaxation oscillators

RELAXATION OSCILLATORS ARE DIFFERENT FROM FEEDBACK OSCILLATORS (WHICH make up the bulk of this book). Recall that feedback oscillators are formed by using a frequency-selective feedback network around a gain amplifier such that the loop gain is greater than unity at the desired operating frequency. Relaxation oscillators, on the other hand, use some device that permits a voltage (usually across a capacitor) to build up and then suddenly drop to a lower level. Two basic forms will be considered here: *neon glow lamp* and *unijunction transistor*.

Neon glow-lamp relaxation oscillators

A neon glow lamp does not use a filament like an incandescent lamp, but rather consists of two electrodes in a glass bulb evacuated of air and then partially refilled with a low-pressure inert noble gas such as neon (Fig. 6-1). At low potentials, the inert gas acts like an insulator, so no current flows in the lamp despite the potential difference across the electrodes. If the potential across the electrodes rises to a certain firing voltage (often around 67 V in common glow lamps), however, the neon gas suddenly ionizes, creating a low electrical resistance and a flash of light. The current flowing in the lamp can be very high under these circumstances, unless it is limited by a series resistance.

The gas inside the lamp remains ionized as the electrode potential drops, until a minimum holding voltage is reached. When this potential is reached, the gas deionizes and the lamp goes out.

Most neon glow lamps (NE-2, NE-51) glow orange. If direct current (dc) is applied, only one electrode inside the glow lamp lights up (which one depends on polarity), but if alternating current (ac) is used, then both electrodes light up.

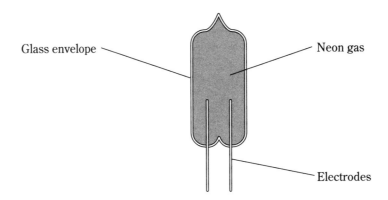

6-1 Neon glow lamp construction.

Experiment 6-1

Connect the circuit of Fig. 6-EXP1. Be extremely cautious. The 90-V battery can be very dangerous. Connect the circuit without the battery, and make certain that switch S1 is turned to the off position (open, as shown). Once the circuit is connected, set potentiometer R1 so that the wiper is close to the grounded end. Connect the 90-V battery, and then (and only then) close S1.

1. Note the voltage reading on M1.
2. Slowly turn R1, noting the rise of the voltage on M1.
3. At some voltage, V, the lamp will ignite. Record that voltage.
4. Now, reverse the direction of R_1, watching the glow lamp. Note the voltage when the lamp extinguishes.

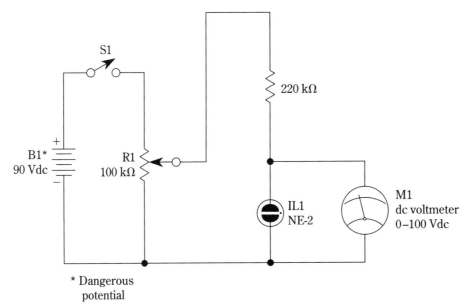

6-EXP1 Circuit for Experiment 6-1.

5. Repeat the experiment several times until you are sure of the values of firing and extinguishing potentials.
6. Turn S1 to the off (open) position.
7. Disconnect the 90-Vdc battery and store it safely so that the electrodes (terminals) are not exposed.

Figure 6-2A shows the circuit of a simple glow-lamp oscillator circuit, while Fig. 6-2B shows the waveforms from turn-on to some time T. The circuit consists of a se-

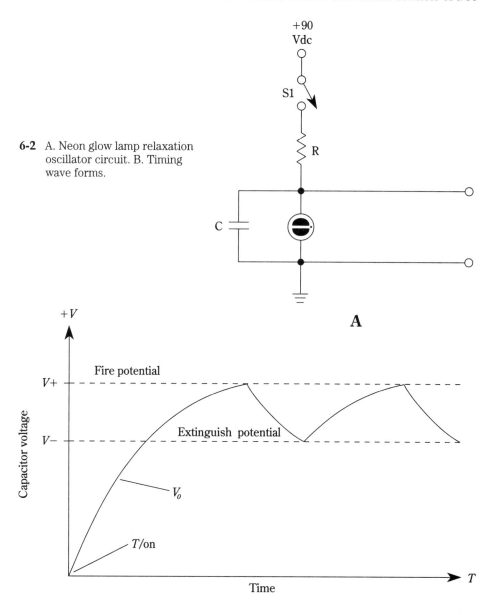

6-2 A. Neon glow lamp relaxation oscillator circuit. B. Timing wave forms.

ries resistance (R), which both limits current flowing in the lamp and provides timing, and a capacitor (C) which finishes the timing. The voltage across the capacitor, which is also the output voltage, follows the rules for resistance-capacitance (RC) networks, as discussed in chapter 9.

The waveform for the oscillator is shown in Fig. 6-2B. At turn-on the voltage across the capacitor (which is also the output voltage, V_o) is zero. But the voltage immediately begins to rise according to:

$$V_o = V (1 - e^{-T/RC}) \tag{6-1}$$

Where: V_o is the voltage across the capacitor
V is the applied voltage (typically 90 Vdc)
T is the elapsed time in seconds
R is the resistance in ohms
C is the capacitance in farads

When the voltage across the capacitor first reaches the firing voltage (V_1), the lamp gas ionizes. This action causes a low resistance to appear across the electrodes, which tends to discharge the capacitor much faster than it charged. The discharge curve from V_2 to V_1 is shown in Fig. 6-2B in exaggerated form in order to show that it has a capacitor-discharge shape. When the output voltage drops to the extinguishing voltage (V_2), the lamp goes out. The resistance across the electrodes is once again high, so the capacitor can begin to charge again. Now the capacitor voltage oscillates back and forth between V_1 and V_2 at a rate determined by the values of R and C.

Experiment 6-2

Connect the circuit of Fig. 6-2A using a 90-V battery and a 680-kΩ, ½-W resistor for R. Initially use a 0.33-μF, 600-WVdc (or higher) capacitor for C. Turn on the switch (S_1) and observe the lamp flashing on and off. Note the number of flashes per minute.

Turn off the switch (open) and disconnect the battery (set aside safely, with the terminals covered). Using an insulated alligator clip lead, connect a short circuit across C. Remove the short and then reconnect it five times. Disconnect the capacitor and select another one in the range of 0.05 to 1 μF. Use only 600-WVdc, or higher, capacitors.

Unijunction transistor relaxation oscillators

The *unijunction transistor*, once also called the *double-base diode*, is a special (and largely archaic) transistor that has two bases (B1 and B2) and an emitter (E). When a dc potential, +V, is applied across B1-B2, no current will flow unless the voltage across the emitter-base-1 terminals is at a certain minimum point. This point is a fraction of the B1-B2 potential, and is usually referred to as the eta (η) voltage.

When the η voltage is exceeded, the resistance of the B1-B2 channel drops very low, and the E-B1 junction breaks down.

Figure 6-3 shows the circuit for a unijunction-transistor (UJT) relaxation oscillator. The timing of the operating frequency is set by R_1 and C_1, while the output voltage is taken across C_1. Opposite polarity pulses are formed at B1 and B2.

At initial turn-on, the voltage across C_1 is zero, but it begins to charge under the influence of +V operating through resistance R_1. The waveform is the same capacitor-charge waveform that we saw in the neon glow-lamp case. When the η voltage is reached, the UJT fires, causing the sudden discharge of the capacitor. This process will continue until turn-off, creating a chain of quasisawtooth waves.

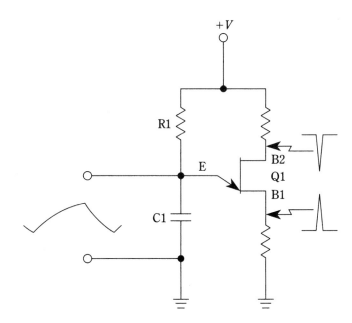

6-3 Unijunction transistor (UJT) relaxation oscillator.

7
Feedback oscillators

A *FEEDBACK OSCILLATOR* (FIG. 7-1) CONSISTS OF AN AMPLIFIER WITH AN OPEN-LOOP gain of A_{vol} and a feedback network with a gain (or transfer function) β. It is called a *feedback oscillator* because the output signal of the amplifier is fed back to the amplifier's own input by way of the feedback network. Figure 7-1 is a block diagram of the feedback oscillator. That it bears more than a superficial resemblance to a feedback amplifier is no coincidence. Indeed, as anyone who has misdesigned or mis-constructed an amplifier knows all too well, a feedback oscillator is an amplifier in which special conditions prevail. These conditions, called *Barkhausen's criteria for oscillation*, include:

- Feedback voltage V_F must be in phase (360°) with the input voltage.
- The loop gain βA_{vol} must be unity (1).

The first of these criteria means that the total phase shift from the input of the amplifier to the output of the amplifier, around the loop back to the input, must be 360° (2ω radians) or an integer (N) multiple of 360° (i.e., N2ω radians).

The amplifier can be any of many different devices. In some circuits it will be a common-emitter bipolar transistor (npn or pnp device). In others it will be a junction field-effect transistor (JFET) or metal-oxide semiconductor field-effect transistor (MOSFET). In older equipment it was a vacuum tube. In modern circuits the active device will probably be either an integrated-circuit operational amplifier or some other form of linear IC amplifier. See Table 7-1.

The amplifier is most frequently an inverting type, so the output is out of phase with the input by 180°. As a result, in order to obtain the required 360° phase shift, an additional phase shift of 180° must be provided in the feedback network at the frequency of oscillation only. If the network is designed to produce this phase shift at only one frequency, then the oscillator will produce a sinewave output on that frequency.

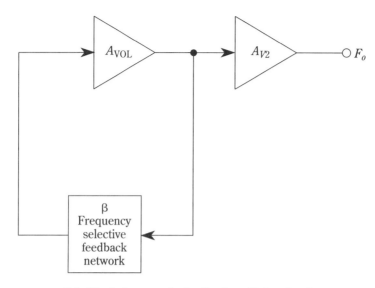

7-1 Block diagram of a feedback oscillator circuit.

Table 7-1

Parameter	Common Emitter	Common Collector	Common Base
Voltage Gain	HIGH	LOW (≈ 1 or less)	HIGH (>100)
Current Gain	HIGH	HIGH (β+1)	LOW (<1)
Z_{in}	Medium (≈ 1)	HIGH (>100K)	LOW
Z_{out}	Medium to High (≈ 50K)	LOW (<100Ω)	HIGH (>500K)
Phase Inversion?	YES	NO	NO

Before considering specific sine wave oscillator circuits, let's examine Fig. 7-1 more closely. Several things can be determined about the circuit:

$$V_i = V_{in} + V_F \tag{7-1}$$

so,

$$V_i = V_{in} - V_F \tag{7-2}$$

and also,

$$V_F = \beta V_o \tag{7-3}$$

$$V_o = V_i A_{vol} \tag{7-4}$$

The transfer function (or gain) A_v is:

$$A_v = \frac{V_o}{V_{in}} \tag{7-5}$$

Substituting Eqs.7-3 and 7-4 into Eq.7-5:

$$A_v = \frac{V_i A_{vol}}{V_i - V_F} \tag{7-6}$$

From Eq. 7-3, $V_F = BV_o$, so:

$$A_v = \frac{V_i A_{vol}}{V_i - \beta V_o} \tag{7-7}$$

However, Eq. 7-4 shows $V_o = V_i A_{vol}$, so Eq. 7-7 can be written:

$$A_v = \frac{V_i A_{vol}}{V_1 - \beta V_i A_{vol}} \tag{7-8}$$

and, dividing both numerator and denominator by V_i:

$$A_v = \frac{A_{vol}}{1 - \beta A_{vol}} \tag{7-9}$$

Equation 7-9 serves for both feedback amplifiers and oscillators. However, in the special case of an oscillator, $V_{in} = 0$, so $V_o \to \infty$. Therefore, it is implied that the denominator of Eq. 7-9 must also be zero:

$$1 - \beta A_{vol} = 0 \tag{7-10}$$

Therefore, for the case of the feedback oscillator:

$$\beta A_{vol} = 1 \tag{7-11}$$

βA_{vol} is the loop gain of the amplifier and feedback network, so Eq. 7–11 meets Barkhausen's second criterion.

8
Resonant RF circuits

IN THIS CHAPTER WE WILL TAKE A LOOK AT INDUCTORS (L) AND CAPACITORS (C), how they are affected by ac signals, and how they are combined into LC tuned circuits. The tuned circuit allows the circuit to be selective about the RF frequency being passed. Alternatively, in the case of oscillators it sets the operating frequency of the circuit.

Tuned resonant circuits

Tuned resonant circuits, also called *tank circuits* or *LC circuits*, are used in a radio front end to select signals from the myriad of signals available at the antenna. The tuned resonant circuit is made up of two principal components: inductors and capacitors. In this section we will examine inductors and capacitors separately and then in combination to determine how they function together to tune the radio's RF, IF, and LO circuits. But first, we need to make a brief digression to discuss vectors, because they are used in describing the behavior of these components and circuits.

Vectors

A *vector* (Fig. 8-1A) is a graphical device that defines the magnitude and direction (both are needed) of a quantity or physical phenomena. The length of the arrow defines the magnitude of the quantity, while the direction in which it is pointing defines the direction of action of the quantity being represented.

Vectors can be used in combination with each other. For example, in Fig. 8-1B we see a pair of displacement vectors that define a starting position (P1) and a final position (P2) for a person who traveled from point P1 12 miles north and then 8 miles east to arrive at point P2. The *displacement* in this system is the hypotenuse of the right triangle formed by the "north" vector and the "east" vector. This concept

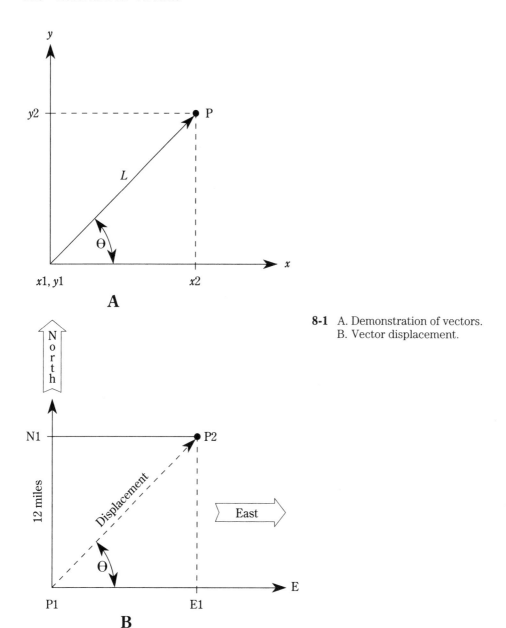

8-1   A. Demonstration of vectors.
B. Vector displacement.

was once illustrated pungently by a university bumper sticker's directions on how to get to a rival school: *"North 'til you smell it, east 'til you step in it."*

The magnitude of the displacement vector to *P2* is given by:

$$P2 = \sqrt{N^2 + E^2} \tag{8-1}$$

However, recall that the magnitude only describes part of the vector's attributes. The other part is the direction of the vector. In the case of Fig. 8-1B, the *direction* can be defined as the angle between the translated "east" vector and the displacement vector. This angle (θ) is given by:

$$\theta = \arccos\left(\frac{E_1}{P}\right) \tag{8-2}$$

In generic vector notation there is no "natural" or "standard" frame of reference, so the vector can be drawn in any direction so long as the users understand what it means. In the system above, we have adopted—by convention—a method that is basically the same as the old fashioned Cartesian coordinate system X-Y graph. In the example of Fig. 8-1B, the X axis is the "east" vector, while the Y axis is the "north" vector.

In electronics, the vectors that describe voltages and currents in ac circuits are standardized (Fig. 8-2) on this same kind of cartesian system in which the inductive reactance (X_L), i.e., the opposition to ac exhibited by inductors, is graphed in the "north" direction, the capacitive reactance (X_c) is graphed in the "south" direction, and the resistance (R) is graphed in the "east" direction. Negative resistance ("west" direction) is sometimes seen in electronics. It is a phenomenon in which the current *decreases* when the voltage increases. RF examples of negative resistance include tunnel diodes and Gunn diodes.

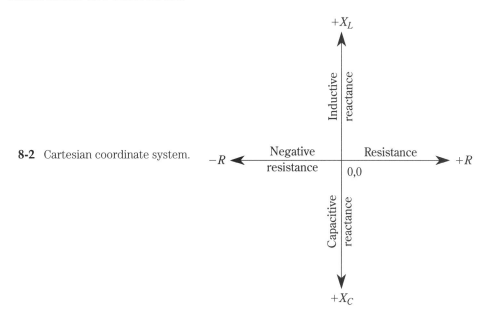

8-2 Cartesian coordinate system.

Inductance and inductors

Inductance (L) is a property of electrical circuits that opposes changes in the flow of current. Note the word "changes"; it is important. As such, it is somewhat analogous to the concept of inertia in mechanics. An inductor stores energy in a magnetic field (a fact which you will see is quite important). In order to understand the concept of inductance, you must understand these physical facts:

- When an electrical conductor moves relative to a magnetic field, a current is generated (or "induced") in the conductor. An *electromotive force* (EMF or "voltage") appears across the ends of the conductor.

- When a conductor is in a magnetic field that is changing, a current is induced in the conductor. As in the first case, an EMF is generated across the conductor.
- When an electrical current moves in a conductor, a magnetic field is set up around the conductor.

According to Lenz's law, the EMF induced into a circuit is ". . . in a direction that opposes the effect that produced it." From this fact we can see the following effects:

- A current induced by either the relative motion of a conductor and a magnetic field, or changes in the magnetic field, always flows in the direction that sets up a magnetic field that opposes the original magnetic field.
- When a current flowing in a conductor changes, the magnetic field that it generates changes in a direction that induces a further current into the conductor that opposes the current change that caused the magnetic field to change.
- The EMF generated by a change in current has a polarity opposite the polarity of the potential that created the original current.

The unit of inductance (L) is the henry (H). The accepted definition of the *henry* is the inductance that creates an EMF of 1 V when the current in the inductor is changing at a rate of one ampere per second, or mathematically:

$$V = L\left(\frac{\Delta I}{\Delta t}\right) \qquad\qquad \textbf{(8-3)}$$

Where: V is the induced EMF in volts
L is the inductance in henrys
I is the current in amperes
t is the time in seconds
Δ indicates a "small change in . . . ."

The henry (H) is the appropriate unit for inductors such as the smoothing filter chokes used in dc power supplies, but is far too large for RF and IF circuits. In those circuits, subunits of millihenrys (mH) and microhenrys (μH) are used. These are related to the henry by: 1 H = 1000 mH = 1,000,000 μH. Thus, 1 mH = 10^{-3} H and 1 μH = 10^{-6} H.

One of the phenomena listed earlier is called *self-inductance*. When the current in a circuit changes, the magnetic field generated by that current change also changes. This changing magnetic field induces a counter current in the direction that opposes the original current change. This induced current also produces an EMF (discussed earlier), which is called the *counter electromotive force* (CEMF). As with other forms of inductance, self-inductance is measured in henrys and its subunits.

Self-inductance can be increased by forming the conductor into a multiple-turn coil (Fig. 8-3) in such a manner that the magnetic field in adjacent turns reinforces each other. This requirement means that the turns of the coil must be insulated from each other. The coil wound in this manner is called an *inductor*, or simply *coil*, in RF/IF circuits. To be grammatically correct, the inductors pictured in Fig. 8-4 are called *solenoid wound coils* if the length (l) is greater than the diameter (d). The inductance of the coil is actually self-inductance, but the "self-" is usually dropped in favor of simply "inductance."

8-3 Air-core inductor.

Several factors affect the inductance of a coil. Perhaps the most obvious are the length, the diameter, and the number of turns in the coil. Also affecting the inductance is the nature of the core material and its cross-sectional area. In the example of Fig. 8-3, the core is simply air and the cross-sectional area is directly related to the diameter; however, in many circuits the core is made of powdered iron or ferrite materials.

For an air-core solenoid wound coil in which the length is greater than 0.4d, the inductance can be approximated by:

$$L_{\mu H} = \frac{d^2 N^2}{18d + 40l} \tag{8-4}$$

The core material has a certain *magnetic permeability* (μ), which is the the ratio of the number of lines of flux produced by the coil with the core inserted to the number of lines of flux with an air core (i.e., core removed). The inductance of the coil is multiplied by the permeability of the core.

Combining inductors When inductors are connected together in a circuit, their inductances combine in a manner similar to the resistances of several resistors in parallel or series. For inductors in which their respective magnetic fields do not interact:

A. Series connected inductors:

$$L_{total} = L_1 + L_2 + L_3 + \ldots + L_n \tag{8-5}$$

B. Parallel connected inductors:

$$L_{total} = \frac{1}{\left(\dfrac{1}{L_1} + \dfrac{1}{L_2} + \dfrac{1}{L_3} + \ldots + \dfrac{1}{L_n} \right)} \tag{8-6}$$

Or, in the special case of two inductors in parallel:

$$L_{total} = \frac{L_1 \times L_2}{L_1 + L_2} \tag{8-7}$$

If the magnetic fields of the inductors in the circuit interact, then the total inductance becomes somewhat more complicated to express. For the simple case of two inductors in series, the expression would be:

A. Series inductors:

$$L_{total} = L_1 + L_2 \pm 2M \qquad \textbf{(8-8)}$$

Where M is the mutual inductance caused by the interaction of the two magnetic fields (note: $+M$ is used when the fields aid each other, and $-M$ is used when the fields are opposing).

B. Parallel inductors:

$$L_{total} = \cfrac{1}{\left(\cfrac{1}{L_1 \pm M}\right) + \left(\cfrac{1}{L_2 \pm M}\right)} \qquad \textbf{(8-9)}$$

Some LC tank circuits use air-core coils in their tuning circuits (Fig. 8-4). Note that two of the coils in Fig. 8-4 are aligned at right angles to the other one. The reason for this arrangement is not mere convenience, but rather is a tactic used by the radio designer in order to prevent interaction of the magnetic fields of the respective coils. In general, for coils in close proximity to each other:

- Maximum interaction between the coils occurs when the coils' axes are parallel to each other.
- Minimum interaction between the coils occurs when the coils' axes are at right angles to each other.

For the case where the coil axes are along the same line, the interaction depends on the distance between the coils.

Adjustable coils   There are several practical problems with the standard fixed coil discussed previously. For one thing, the inductance cannot easily be adjusted either to tune the radio or to trim the tuning circuits to account for the tolerances in the circuit.

Air-core coils are difficult to adjust. They can be lengthened or shortened; the number of turns can be changed; or a tap or series of taps can be established on the coil in order to allow an external switch to select the number of turns that are al-

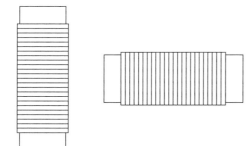

8-4   Solenoid wound inductors at right angles to each other.

lowed to be effective. None of these methods is terribly elegant, even though all have been used in one application or another.

The solution to the adjustable-inductor problem developed relatively early in the history of mass-produced radios, and still used today, is to insert a powdered-iron or ferrite core (or "slug") inside the coil form (Fig. 8-5). The permeability of the core increases or decreases the inductance according to how much of the core is inside the coil. If the core is made with either a hexagonal hole or screwdriver slot, then the inductance of the coil can be adjusted by moving the core in or out of the coil. These coils are called *slug-tuned inductors*.

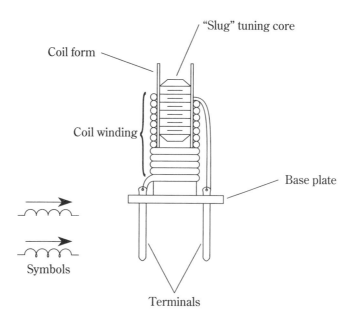

8-5 Slug-tuned coil.

Inductors in ac circuits

Impedance (Z) is the total opposition to the flow of alternating current (ac) in a circuit, and as such it is analogous to resistance in dc circuits. The impedance is made up of a resistance component (R) and a component called reactance (X). Like resistance, reactance is measured in ohms. If the reactance is produced by an inductor, then it is called *inductive reactance* (X_L), and if by a capacitor it is called *capacitive reactance* (X_c). Inductive reactance is a function of the inductance and the frequency of the ac source:

$$X_L = 2\pi f L \qquad \text{(8-10)}$$

Where: X_L is the inductive reactance in ohms
f is the ac frequency in hertz (Hz)
L is the inductance in henrys (H)

In a purely resistive ac circuit (Fig. 8-6A) the current (I) and voltage (V) are said to be *in-phase* with each other; i.e., they rise and fall at exactly the same times in the ac cycle. In vector notation (Fig. 8-6B), the current and voltage vectors are along the same axis, which is an indication of the zero-degree phase difference between the two.

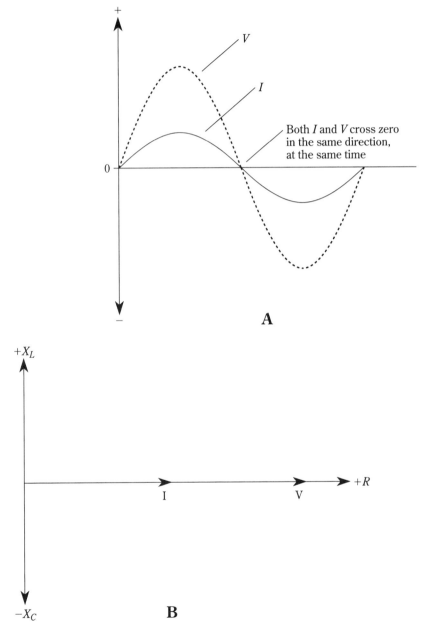

Both I and V cross zero in the same direction, at the same time

A

B

8-6 A. Voltage and current in phase with each other. B. Vector representation.

In an ac circuit that contains only an inductor (Fig. 8-7A) and is excited by a sine-wave ac source, change in current is opposed by the inductance. As a result, the current (I) in an inductive circuit lags behind the voltage (V) by 90°. This is shown vectorially in Fig. 8-7B, and as a pair of sine waves in Fig. 8-7C.

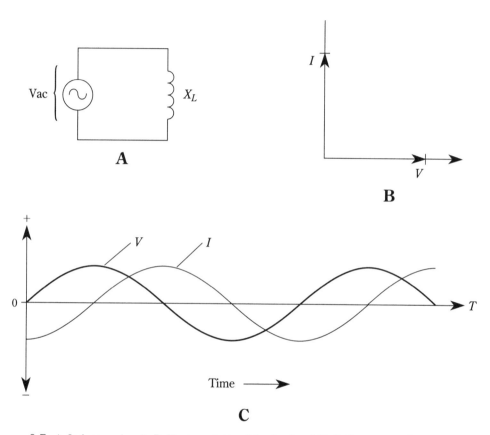

8-7 A. Inductor circuit. B. Vector relationships for I and V. C. Sine-wave relationships.

The ac circuit that contains a resistance and an inductance (Fig. 8-8A) shows a phase shift (θ), shown vectorially in Fig. 8-8B, that is other than the 90° seen in purely inductive circuits. The phase shift is proportional to the voltage across the inductor and the current flowing through it. The impedance (Z) of this circuit is found by the Pythagorean rule described earlier, also called the *root of the sum of the squares* method (see Fig. 8-8):

$$Z = \sqrt{R^2 X_L)^2} \qquad\qquad \textbf{(8-11)}$$

The coils used in radio receivers come in a variety of different forms and types, but all radios (except the very crudest untuned crystal sets) will have at least one coil. Now turn your attention to the other member of the LC tuned circuit, the capacitor.

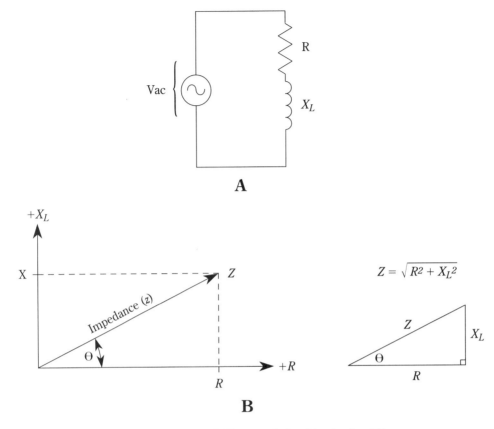

$$Z = \sqrt{R^2 + X_L{}^2}$$

8-8   A. RL circuit. B. Vector relationships for *I* and *V*.

Capacitors and capacitance

Capacitors (also called condensers in early texts) are the other component used in radio tuning circuits. Like the inductor, the capacitor is an energy-storage device. While the inductor stores electrical energy in a magnetic field, the capacitor stores energy in an electrical (or electrostatic) field. Electrical charge (Q) is stored in the capacitor.

The basic capacitor consists of a pair of metallic plates facing each other, and separated by an insulating material called a *dielectric*. This arrangement is shown schematically in Fig. 8-9A, and in a more physical sense in Fig. 8-9B. The fixed capacitor shown in Fig. 8-9B consists of a pair of square metal plates separated by a dielectric. Although this type of capacitor is not terribly practical, it was once used quite a bit in transmitters. Spark transmitters of the 1920s often used a glass and tinfoil capacitor fashioned very much like Fig. 8-9B. Layers of glass and foil are sandwiched together to form a high-voltage capacitor. A 1-foot-square capacitor made of ⅛-inch-thick glass and foil has a capacitance of about 2000 pF.

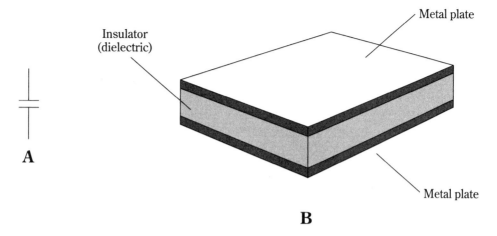

8-9 A. Capacitor circuit symbol. B. "Glass plate" capacitor.

Units of capacitance

The *capacitance* (*C*) of the capacitor is a measure of its ability to store current, or more properly, electrical charge. The principal unit of capacitance is the farad (named after physicist Michael Faraday). One *farad* is the capacitance that will store one coulomb of electrical charge (6.28×10^{18} electrons) at an electrical potential of 1 V. Or, in math form:

$$C_{farads} = \frac{Q_{coulombs}}{V_{volts}} \qquad (8\text{-}12)$$

The farad is far too large for practical electronics work, so subunits are used. The microfarad (μF) is 0.000001 *F* (1 F = 10^6 μF). The picofarad (pF) is 0.000001 μF, or 10^{-12} F. In older radio texts and schematics the picofarad was called the micromicrofarad ($\mu\mu F$), but never fear: 1 $\mu\mu F$ = 1 pF.

The capacitance of the capacitor is directly proportional to the area of the plates (in terms of Fig. 8-9B, L × W), inversely proportional to the thickness (*T*) of the dielectric (or the spacing between the plates, if you prefer), and directly proportional to the dielectric constant (*k*) of the dielectric.

The dielectric constant is a property of the insulator material used for the dielectric. The dielectric constant is a measure of the material's ability to support electric flux, and is thus analogous to the permeability of a magnetic material. The standard of reference for dielectric constant is a perfect vacuum, which is said to have a value of *k* = 1.00000. Other materials are compared with the vacuum. The values of *k* for some common materials are:

Vacuum	1.0000
Dry air	1.0006
Paraffin (wax) paper	3.5
Glass	5 to 10
Mica	3 to 6

Rubber	2.5 to 35
Dry wood	2.5 to 8
Pure (distilled) water	81

The value of capacitance in any given capacitor is found from:

$$C = \frac{0.0885kA\ (N{-}1)}{T} \tag{8-13}$$

Where: C is the capacitance in picofarads (pF)

k is the dielectric constant

A is the area of one of the plates $(L \times W)$, assuming that the two plates are identical

N is the number of identical plates

T is the thickness of the dielectric

Breakdown voltage

The capacitor works by supporting an electrical field between two metal plates. This potential, however, can get too large. When the electrical potential, i.e., the voltage, gets too large, free electrons in the dielectric material (there are a few, but not many, in any insulator) may flow. If a stream of electrons gets started, then the dielectric might break down and allow a current to pass between the plates. The capacitor is then said to be *shorted* (heck, it's not *said* to be shorted, *it is shorted!*). The maximum breakdown voltage of the capacitor must not be exceeded. However, for practical purposes there is a smaller voltage called the *dc working voltage* (WVdc) rating that defines the maximum safe voltage that can be applied to the capacitor. Typical values found in common electronic circuits vary from 8 WVdc to 1000 WVdc.

Circuit symbols for capacitors

The circuit symbols used to designate fixed-value capacitors are shown in Fig. 8-10A, and for variable capacitors in Fig. 8-10B. Both types of symbol are common. In certain types of capacitor, the curved plate shown on the left in Fig. 8-10A is usually the outer plate, i.e., the one closest to the outside package of the capacitor. This end of the capacitor is often indicated with a color band next to the lead attached to that plate.

The symbols for the variable capacitor are shown in Fig. 8-10B. This symbol is the fixed-value symbol with an arrow through the plates. Small trimmer and padder capacitors are often denoted by the symbol of Fig. 8-10C. The variable set of plates is designated by the arrow.

Fixed capacitors

Several types of fixed capacitors are found in typical electronic circuits, and these are classified by dielectric type: paper, mylar, ceramic, mica, polyester, and others.

The construction of old-fashioned paper capacitors is shown in Fig. 8-11. It consists of two strips of metal foil sandwiched on both sides of a strip of paraffin-wax paper. The strip sandwich is then rolled-up into a tight cylinder. This rolled-up cylinder is then packaged in either a hard plastic, bakelite, or paper-and-wax case. When the case is cracked, or the wax end plugs are loose, replace the capacitor even though it

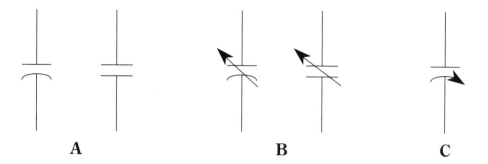

8-10 Capacitor circuit symbols: A. Fixed capacitors. B. Variable capacitors. C. Trimmer (variable) capacitors.

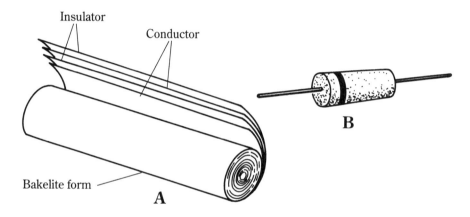

8-11 Paper capacitor. A. Internal structure. B. Finished capacitor.

tests good—it won't be for long. Paper capacitors come in values from about 300 pF to about 4 μF. The breakdown voltages will be 100 WVdc to 600 WVdc.

The paper capacitor is used for a number of different applications in older circuits, such as bypassing, coupling, and dc blocking. Unfortunately, no component is perfect. The long rolls of foil used in the paper capacitor exhibit a significant amount of stray inductance. As a result, the paper capacitor is not used for high frequencies. Although they are found in some shortwave receiver circuits, they are rarely or never used at VHF.

In modern applications, or when servicing older equipment that used paper capacitors, use a mylar dielectric capacitor in place of the paper capacitor. Select a unit with exactly the same capacitance rating, and a WVdc rating that is equal to or greater than the original WVdc rating.

Several different forms of ceramic capacitors are shown in Fig. 8-12. These capacitors come in values from a few picofarads up to 0.5 μF. The working voltages range from 400 WVdc to more than 30,000 WVdc. The common garden-variety disk

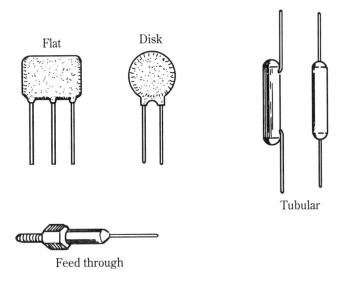

Flat

Disk

Tubular

Feed through

8-12   Ceramic capacitors.

ceramic capacitors are usually rated at either 600 WVdc or 1000 WVdc. Tubular ceramic capacitors are typically much smaller in value than disk or flat capacitors, and are used extensively in VHF and UHF circuits for blocking, decoupling, bypassing, coupling, and tuning.

The feedthrough type of ceramic capacitor is used to pass dc and low-frequency ac lines through a shielded panel. These capacitors are often used to filter or decouple lines that run between circuits that are separated by a shield for reducing electromagnetic interference (EMI).

Ceramic capacitors are often rated as to *temperature coefficient*. This specification is the change of capacitance per change of temperature in degrees Celcius. A "P" prefix indicates a positive temperature coefficient, an "N" indicates a negative temperature coefficient, and the letters "NPO" indicate a zero temperature coefficient (*NPO* stands for "negative positive zero"). Do not ad-lib on these ratings when servicing a piece of electronic equipment. Use exactly the same temperature coefficient as the original manufacturer used. Nonzero temperature coefficients are often used in oscillator circuits to temperature-compensate the oscillator's frequency drift.

Several different types of mica capacitor are shown in Fig. 8-13. The fixed mica capacitor consists of either metal plates on either side of a sheet of mica, or a sheet of mica that is silvered with a deposit of metal cn either side. The range of values for mica capacitors tends to be 50 pF to 0.02 μF at voltages in the range of 400 WVdc to 1000 WVdc. The mica capacitor shown in Fig. 8-13C is called a *silvered mica* capacitor. These capacitors are low temperature coefficient, although for most applications an NPO disk ceramic will service better than all but the best silvered mica units. Mica capacitors are typically used for tuning and other uses in higher frequency applications.

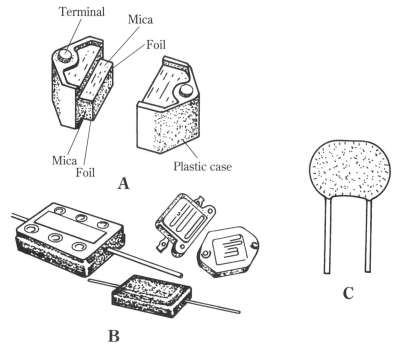

8-13 Mica capacitors.

Other capacitors   Today the equipment designer has a number of different dielectric capacitors available that were not commonly available (or available at all) a few years ago. Polycarbonate, polyester, and polyethylene capacitors are used in a wide variety of applications where the previously discussed capacitors once ruled supreme. In digital circuits we find tiny 100-WVdc capacitors with values of 0.01 μF to 0.1 μF. These are used for decoupling noise on the +5 Vdc power-supply line. In circuits such as timers and op-amp Miller integrators, where the leakage resistance across the capacitor becomes terribly important, you might want to use a polyethylene capacitor. Check current catalogs for various old and new style capacitors. The applications paragraph in the catalog tells you in which applications they will serve; that is a guide to the types of antique capacitor they replace.

Capacitors in ac circuits

When an electrical potential is applied across a capacitor, current will flow as charge is stored in the capacitor. As the charge in the capacitor increases, the voltage across the capacitor plates rises until it equals the applied potential. At this point the capacitor is fully charged, and no further current will flow.

Figure 8-14 shows an analogy for the capacitor in an ac circuit. The actual circuit is shown in Fig. 8-14A, and consists of an ac source connected in parallel across the capacitor (*C*). The mechanical analogy is shown in Fig. 8-14B. The "capacitor" (*C*) consists of a two-chamber cylinder in which the upper and lower chambers are separated by a flexible membrane or diaphragm. The "wires" are pipes to the "ac

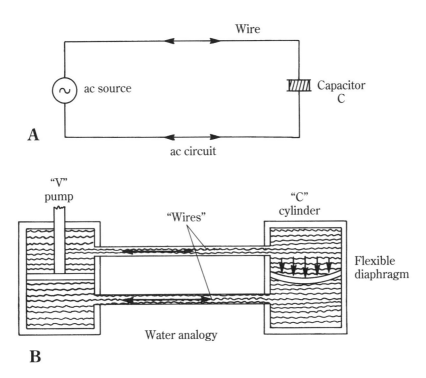

8-14 A. Capacitor ac circuit. B. Water system analogy.

source" (which is a pump). As the pump moves up and down, pressure is applied to first one side of the diaphragm and then the other, alternately forcing fluid to flow into and out of the two chambers of the "capacitor."

The ac-circuit mechanical analogy is not perfect, but it works for our purposes. Now let's apply these ideas to the electrical case. Figure 8-15 shows a capacitor connected across an ac (sine-wave) source. In Fig. 8-15A, the ac source is positive, so negatively charged electrons are attracted from plate A to the ac source, and electrons from the negative terminal of the source are repelled towards plate B of the capacitor. On the alternate half-cycle (Fig. 8-15B), the polarity is reversed, so electrons from the new negative pole of the source are repelled toward plate A of the capacitor, and electrons from plate B are attracted toward the source. Thus, current will flow in and out of the capacitor on alternating half cycles of the ac source.

Voltage and current in capacitor circuits Consider the circuit in Fig. 8-16: an ac source (V) connected in parallel with the capacitor (C). It is the nature of a capacitor to oppose these changes in the applied voltage (the inverse of the action of an inductor). As a result, the voltage (V) lags behind the current (I) by 90°. These relationships are shown in terms of sine waves in Fig. 8-16B, and in vector form in Fig. 8-16C.

Do you want to remember the difference between the action of inductors (L) and capacitors (C) on the voltage and current? Earlier texts used the letter E to denote voltage, so could make a little mnemonic:

ELI the ICE man

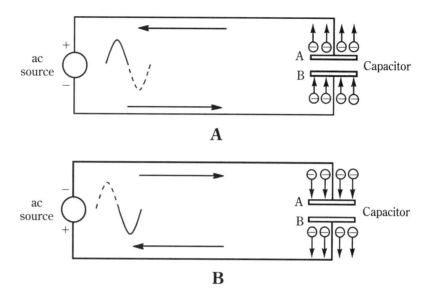

8-15 Ac circuit action on two alternate half cycles.

"ELI the ICE man" suggests that in the inductive (L) circuit, the voltage (E) comes before the current (I)—ELI, and in a capacitive (C) circuit the current comes before the voltage—ICE.

The action of a circuit containing a resistance and capacitance is shown in Fig. 8-17A. As in the case of the inductive circuit, there is no phase shift across the resistor, so the R vector points in the "east" direction (Fig. 8-17B). The voltage across

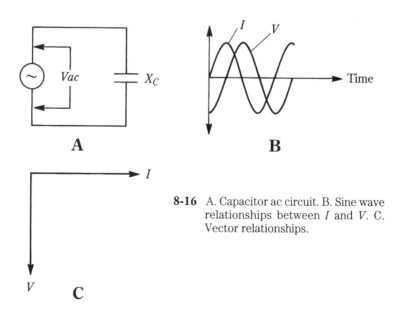

8-16 A. Capacitor ac circuit. B. Sine wave relationships between I and V. C. Vector relationships.

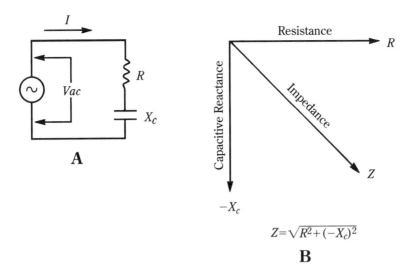

8-17 A. RC ac network. B. Vector relationships.

the capacitor, however, is phase-shifted −90°, so its vector points "south." The total resultant phase shift (**θ**) is found using the Pythagorean rule to calculate the angle between V_r and V_t.

The impedance of the RC circuit is found in exactly the same manner as the impedance of an RL circuit, i.e., the root of the sum of the squares:

$$Z = \sqrt{R^2 + (X_c)^2} \tag{8-14}$$

LC resonant tank circuits

When you use an inductor (L) and a capacitor (C) together in the same circuit, the combination forms an *LC resonant circuit*, also sometimes called a *tank circuit* or *resonant tank circuit*. These circuits are used to tune a radio receiver. There are two basic forms of LC resonant tank circuit: *series* (Fig. 8-18A) and *parallel* (Fig. 8-18B). These circuits have much in common and much that makes them fundamentally different from each other.

The condition of resonance occurs when the capacitive reactance (X_c) and inductive reactance (X_L) are equal. As a result, the resonant tank circuit shows up as purely resistive at the resonant frequency (see Fig. 8-18C) and as a complex impedance at other frequencies. The LC resonant tank circuit operates by an oscillatory exchange of energy between the magnetic field of the inductor and the electrostatic field of the capacitor, with a current between them carrying the charge.

Because the two reactances are both frequency dependent and because they are inverse to each other, the resonance occurs at only one frequency (f_r). We can calculate the standard resonance frequency by setting the two reactances equal to each other and solving for f. The result is:

$$f = \frac{1}{2\pi\sqrt{LC}} \tag{8-15}$$

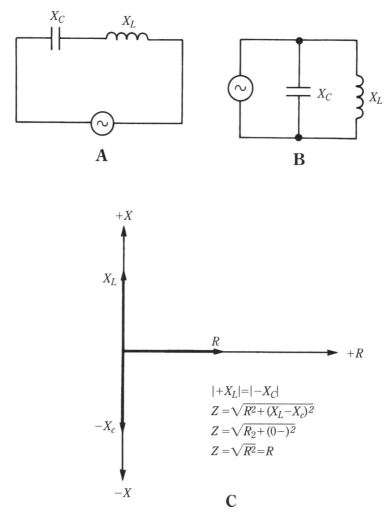

8-18 A. LC series-parallel-resonant circuit. B. LC parallel-resonant circuit. C. Vector relationships show $X_L = X_c$ at resonance.

Series-resonant circuits The series-resonant circuit (Fig. 8-18A), like other series circuits, is arranged so that the terminal current (I) from the source (V) flows in both components equally. The vector diagrams of Fig. 8-19A through 8-19C show the situation under three different conditions.

In the first condition (Fig. 8-19A), the inductive reactance is larger than the capacitive reactance, so the excitation frequency is greater than f_r. Note that the voltage drop across the inductor is greater than that across the capacitor, so the total circuit looks like it contains a small inductive reactance.

In the second condition (Fig. 8-19B), the situation is reversed—the excitation frequency is less than the resonant frequency, so the circuit looks slightly capacitive to the outside world.

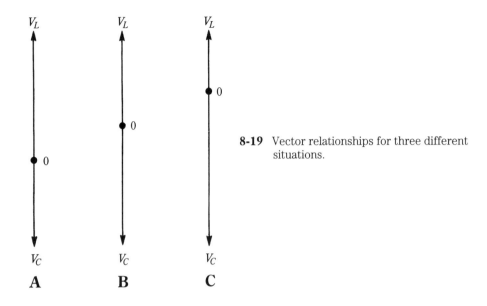

8-19   Vector relationships for three different situations.

In the third condition (Fig. 8-19C), the excitation frequency is at the resonant frequency, so $X_c = X_L$ and the voltage drops across the two components are equal but of opposite phase.

In a circuit that contains a resistance, an inductive reactance, and a capacitive reactance, there are three vectors (X_L, X_c, and R) to consider (Fig. 8-20), plus a re-

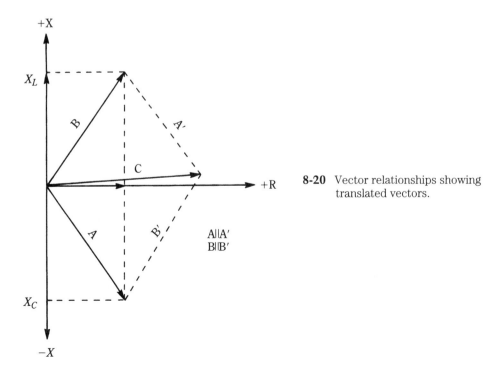

8-20   Vector relationships showing translated vectors.

sultant vector. As in the other circuit, the "north" direction represents X_L, the "south" direction represents X_C, and the "east" direction represents R. Using the parallelogram method, we first construct a resultant for the R and X_C, which is shown as vector "A". Next, we construct the same kind of vector ("B") for R and X_L. The resultant ("C") is made using the parallelogram method on "A" and "B". Vector "C" represents the impedance of the circuit; the magnitude is represented by the length, and the phase angle by the angle between "C" and R.

Figure 8-21A shows a series-resonant LC tank circuit, and Fig. 8-21B shows the current and impedance as a function of frequency.

A series-resonant circuit has a low impedance at its resonant frequency and a high impedance at all other frequencies. As a result, the line current (I) from the source is maximum at the resonant frequency and the voltage across the source is minimum.

Parallel-resonant circuits The parallel-resonant tank circuit (Fig. 8-22A) is the inverse of the series-resonant circuit. The line current (I) from the source splits and flows in the inductor and the capacitor separately.

The parallel-resonant circuit has its highest impedance at the resonant frequency and a low impedance at all other frequencies. Thus, the line current from the source is minimum at the resonant frequency (Fig. 8-22B), and the voltage across the LC tank circuit is maximum. This fact is important in radio tuning circuits.

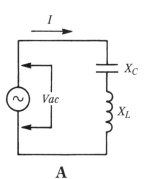

A

8-21 A. Series resonant circuit.
B. Current and voltage vs frequency graph.

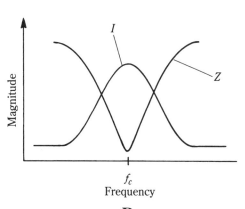

B

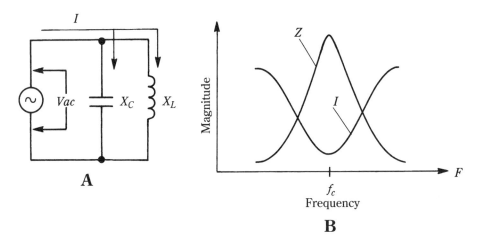

8-22 A. Parallel resonant circuit. B. Current and voltage vs frequency graph.

9

RC networks
The key to timing
in multivibrator circuits

A LARGE NUMBER OF THE WAVEFORM-GENERATOR AND OSCILLATOR CIRCUITS in this book depend on the characteristics of the simple resistor-capacitor (RC) network for their proper operation. Indeed, many of the experiments are based on the behavior of these two simple components. Understanding RC networks is key to understanding the circuits that use them. Unfortunately, many books on RC-based oscillators overlook the characteristics of these circuits. In order to overcome this shortcoming, this chapter is provided as a brief review of RC-network dc circuit theory.

Consider Fig. 9-1A. Assuming that the initial condition is as shown, switch S1 is in position A and is thus open circuited. There is initially no electrical charge stored in capacitor C (i.e., $V_c = 0$). If switch S1 is moved to position B, however, voltage V is applied to the RC network. The capacitor begins to charge with current from the battery, and V_c begins to rise towards V (see curve V_{cb} in Fig. 9-1B). The instantaneous capacitor voltage is found from:

$$V_c = V(1 - e^{-T/RC}) \qquad (9\text{-}1)$$

Where: V_c is the capacitor voltage
$\quad V$ is the applied voltage from the source
$\quad T$ is the elapsed time (in seconds) after charging begins
$\quad R$ is the resistance in ohms
$\quad C$ is the capacitance in farads

The product RC is called the *time constant* of the network, and is sometimes abbreviated τ. If R is in ohms and C is in farads, then the product RC is in seconds. The capacitor voltage rises to approximately 63.2 percent of the final "fully charged" value after 1RC, 86 percent after 2RC and >99 percent after 5RC. A capacitor in an RC network is considered "fully charged" by definition after five time constants (>99.97%).

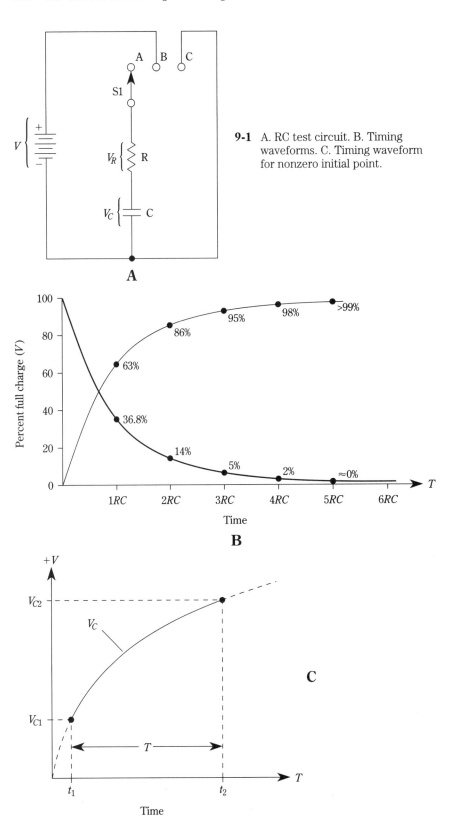

9-1   A. RC test circuit. B. Timing waveforms. C. Timing waveform for nonzero initial point.

If switch S1 in Fig. 9-1A is next set to position C, the capacitor will begin to discharge through the resistor. In the discharge condition:

$$V_C = Ve^{-T/RC} \qquad\qquad \textbf{(9-2)}$$

Voltage V_c drops to 36.8 percent of the full charge level after one time constant ($1RC$), and to very nearly asymptotic to zero after $5RC$. A capacitor in an RC network is considered fully discharged after five time constants (see safety note at end of chapter).

Next consider Fig. 9-1C. This graph represents a situation commonly encountered in waveform-generator circuits. In this graph the capacitor is required to charge from some initial condition (V_{C1}), which may or may not be 0 V, to a final condition (V_{C2}), which may or may not be the fully charged "$5RC$" point, in a specified time interval, T. The question asked by a circuit designer is: "What RC time constant will force V_{C1} to rise to V_{C2} in time T? Assuming that $V_{C1} < V_{C2} < V$:

$$V - V_{C2} = (V - V_{C1})\, e^{-T/RC} \qquad\qquad \textbf{(9-3)}$$

$$\frac{V - V_{C2}}{V - V_{C1}} = e^{-T/RC} \qquad\qquad \textbf{(9-4)}$$

or, rearranging terms:

$$RC = \frac{-T}{\ln\left(\dfrac{V - V_{C2}}{V - V_{C1}}\right)} \qquad\qquad \textbf{(9-5)}$$

Example 9-1

An RC network is connected to a +12 Vdc source. What RC product will permit voltage V_c to rise from +1 Vdc to +4 Vdc in 200 ms? NOTE: V = +12 Vdc, V_{C2} = +4 Vdc, and V_{C1} = +1 Vdc.

Solution:

$$
\begin{aligned}
RC &= \frac{-T}{\ln\left(\dfrac{V - V_{C2}}{V - V_{C1}}\right)} \qquad\qquad \textbf{(9-6)}\\[2em]
&= \frac{-\left(200\ \text{ms}\left(\dfrac{1\ \text{s}}{1000\ \text{ms}}\right)\right)}{\ln\left(\dfrac{12 - 4}{12 - 1}\right)}\\[2em]
&= \frac{-0.200\ \text{s}}{\ln\left(\dfrac{8}{11}\right)}
\end{aligned}
$$

$$= \frac{-0.200 \text{ s}}{\ln(0.727)} \qquad \textbf{(9-6 Cont.)}$$

$$= \frac{(-0.200 \text{ s})}{(-0.319)}$$

$$= 0.627 \text{ s}$$

Equation 9-3 can be used to derive the timing or frequency-setting equations of most RC-based nonsinusoidal waveform-generator or oscillator circuits. The key voltage levels will, most often, be device trip points or critical values set by the design of the circuit.

Experiment 9-1

This experiment is used to explore the capacitor-charging properties of RC networks. (See Fig. 9-2.) A low-voltage source (9-V battery or equivalent dc power supply) is used to charge a 100-µF capacitor. Select a capacitor with a voltage rating of 15 WVdc or higher. Select three fixed resistors of 10 kΩ, 100 kΩ, and 1 MΩ. A 0 to 10 Vdc voltmeter is used to measure the voltage across the capacitor (V_c).

1. Measure the value of R_1, and record the results.
2. Likewise, measure the values of R_2 and R_3, and record the results.
3. Calculate the following time constants:

$$RC_1 = R_1C_1 \text{ s}$$
$$RC_2 = R_2C_1 \text{ s}$$
$$RC_3 = R_3C_1 \text{ s}$$

4. Connect the components into the circuit. Use push-button SPST switches for S2 and S3, and a single-pole, three-position (SP3P) slide or rotary switch for S1.
5. Press S3 several times to ensure that capacitor C_1 is discharged.
6. Record V_{C1} (it should be zero).
7. Set switch S1 to position "A."
8. While watching the dc voltmeter, press switch S2 and hold until there is no further increase in V_{C1}, and then release S2. How fast was it? Nominally, it would be 1 s, which is a little fast to see. However, at longer time constants you will be able to measure the time with a watch second hand.
9. Note the reading on the meter. If the capacitor is ideal (it isn't), the voltage will remain at the final value for an indefinite period. In real circuits, stray "leakage" resistances will bleed off the charge in a relatively short time. How does the final value of V_{C1} compare with the battery voltage of B1?
10. Next, press S3 several times to ensure that the capacitor voltage returns to zero.
11. Use a watch second hand to record the charging times in the next two readings.
12. Set switch S1 to position "B."
13. Press switch S2, and hold it until there is no further increase in V_{C1}, and then release S2. How long did it take to reach the final value? Is this final value similar to the final voltage in the above step? How does the time required to charge the capacitor compare with the calculated RC time constant RC_2?

14. Now, discharge C_1 by pressing and holding S3 several times.
15. Move switch S1 to position "C."
16. Press S2 and hold it down until there is no further increase in the voltage across C_1, and then release S2. Note the voltage across C_1. How long did it take to reach this voltage? How does that time compare with RC_3?
17. It is likely that your calculations and observations differed by a small amount. Part of this error is due to the fact that the actual value of C_1 is not 100 µF, but rather close to that value plus or minus a small tolerance. Also, the accuracy of your time measurements is not perfect.

This experiment suggests a method for measuring the capacitance of a test capacitor. By measuring the resistance and battery voltage very accurately, and noting the time required to reach a fully charged value (or some other point), the capacitance can be calculated.

Trip-point circuits

A *trip-point circuit* is a circuit that remains in one state until a certain input voltage level is reached, and then abruptly switches to an alternate state. Trip-point circuits are at the heart of certain waveform-generator circuits studied in subsequent chapters of this book.

The operational-amplifier voltage-comparator circuit is an example of a trip-point circuit. A voltage comparator compares two input voltages (V_1 and V_2), and issues an output to indicate which relationship is true: $V_1 = V_2$, $V_1 < V_2$, or $V_1 > V_2$. Figure 9-2 shows a voltage comparator connected as a trip-point circuit. In this circuit, the noninverting input of the op amp is biased to V_2 by a zener diode voltage regulator. The zener diode potential, V_Z, is V_2.

Voltage V_1 is the voltage V_{C1} across capacitor C_1. When the circuit is first turned on, V_2 rapidly goes to V_Z, but V_1 is zero. In this case, $V_1 < V_2$, so the output voltage V_o is high (positive).

Capacitor C_1 begins charging through R_1 immediately, however, so V_{C1} begins rising. At some point, $V_1 = V_2$, so the output V_o drops to zero, but does not remain there long. Almost immediately, voltage $V_1 > V_2$, so the output snaps from zero to low (negative). From the observer's point of view, the switch from high to low (i.e., positive to negative) is almost immediate.

Experiment 9-2

Connect the circuit of Fig. 9-2. Monitor the output either with a voltage meter or the light-emitting-diode (LED) circuit shown in Fig. 9-2. In either case, monitor the voltage across C_1 with a voltmeter.

1. Turn on the circuit and let the output stabilize to $V_1 > V_2$. One LED (D1) should be turned on, and the other turned off. Measure V_2.
2. Press S1 and hold it down until V_{C1} is zero.
3. Release S1, while watching both the LEDs and the voltmeter. Note the voltage at which one LED turns off and the other turns on.

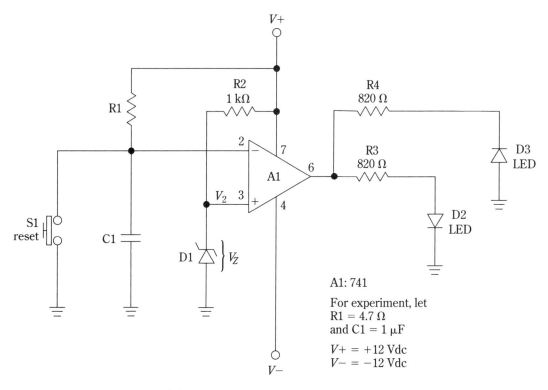

9-2 Circuit for Experiments 9-1 and 9-2.

Note that this time is predictable, and repeatable. Therefore, combining a trip-point circuit, such as a voltage comparator, and an RC network makes it possible to create a timing circuit for a waveform generator.

Safety note

High-voltage capacitors can be dangerous. Although none of the experiments and projects in this book deal with high-voltage capacitors, you might acquire them at some point in your electronics careers. Therefore, a note of caution is in order.

Some of the capacitors found in the ripple filters of high-voltage dc power supplies can contain a high voltage charge, with sufficient stored current to kill humans. Do not rely on the 5RC rule in safety situations. When high voltages are used, some residual electrical energy will be stored in the capacitor dielectric after discharge. This charge can be dangerous. Repetitive discharges should be performed in order to reduce this "residual" charge to zero before handling it. If you are unsure of yourself in dealing with high-voltage capacitors, then refer servicing and handling to a qualified person. This phenomenon is less a factor in low-voltage capacitors, but is a hazard in high-voltage types. Be careful.

Now that we've taken a look at basic RC networks and trip-point circuits, let's start our study of oscillators, multivibrators, and other waveform-generator circuits.

10

The effect of RLC circuits on electronic waveforms

MANY OF THE OSCILLATORS AND MULTIVIBRATORS DISCUSSED IN THIS BOOK CAN use some form of waveshaping circuit, in the form of RC and RLC networks, to alter the shape of an oscillator output signal. In this chapter we will look at some of the reasons why RC and RLC circuits can accomplish this neat trick.

All continuous periodic signals can be represented by a fundamental frequency sine wave (f), and a collection of harmonics (nf) of that fundamental sine wave, that are summed together linearly. These frequencies comprise the Fourier series of the waveform. The elementary sine wave is described by:

$$v = V_m \sin(2 \, \omega \, t) \tag{10-1}$$

Where: v is the instantaneous amplitude of the sine wave at time t
V_m is the peak amplitude of the sine wave
ω is the angular frequency ($2\pi f$) of the sine wave
t is the time in seconds

The *period* of the sine wave is the time between recurrence of identical events, or $T = 2\pi/\omega = 1/f$, where f is the frequency in cycles per second (see Fig. 10-1).

Figure 10-2 shows three complex waveforms and the sine and cosine waves that make them up. Figure 10-2A shows the basic symmetrical square wave. In Fig. 10-2B, the harmonic (curve "A") sine wave of frequency f is added to its third harmonic (curve "B"), at frequency $3f$, to produce a distorted square wave shown as Curve "C." For comparison, the idealized square wave is shown superimposed on the group of waves. In Figs. 10-2C and 10-2D, the process continues with harmonics up to the seventh ($7f$); notice that the square wave is becoming more like the underlying ideal square wave. The construction of the sawtooth and peaked near-triangle waveforms are shown in Figs. 10-3 and 10-4, respectively.

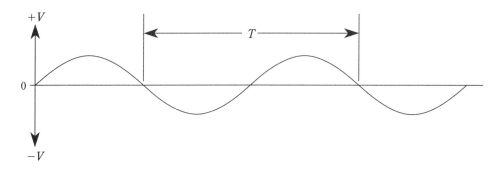

10-1   Sine wave showing one complete cycle of period T.

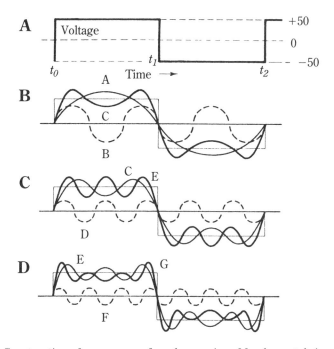

10-2   Construction of square wave from harmonics of fundamental sine wave.

The Fourier series that makes up a waveform can be found if a given waveform is decomposed into its constituent frequencies either by a bank of harmonically re-lated frequency-selective filters, or in a computer using a digital signal-processing al-gorithm called the *Fast Fourier Transform* (FFT).

The Fourier series needed to construct a particular waveform from the ground up can be calculated mathematically. In some of the math used below, the notation of summation and integral calculus is used, but you don't need to know any ad-vanced mathematics in order to understand these concepts. In fact, after you read the passage, don't worry about it again.

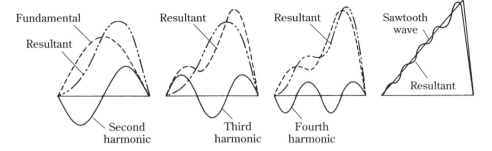

10-3 Construction of sawtooth from a fundamental sine wave and its harmonics.

10-4 Peaked waveform constructed of a fundamental sine wave and its harmonics.

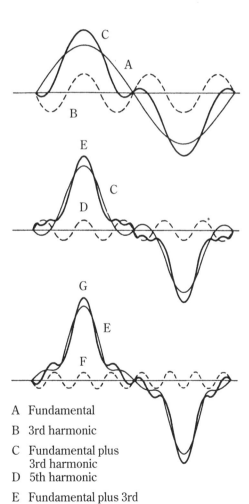

A Fundamental

B 3rd harmonic

C Fundamental plus 3rd harmonic

D 5th harmonic

E Fundamental plus 3rd and 5th harmonics

F 7th harmonic

G Fundamental plus 3rd, 5th and 7th harmonics

The symbol "Σ" means "summation." For example,

$$Y = \sum_{n=1}^{5} X_n \tag{10-2}$$

means to add together all values of X_n together over the range $n = 1$ to $n = 5$. For example, suppose the date set is: $X_1 = 2.5$, $X_2 = 0$, $X_3 = 1.7$, $X_4 = 2.8$, and $X_5 = 3.4$. It would then be applied as follows:

$$Y = \sum_{n=1}^{5} X_n \tag{10-3}$$

$$= X_1 + X_2 + X_3 + X_4 + X_5$$
$$= 2.5 + 0 + 1.7 + 2.8 + 3.4$$
$$= 10.4$$

The only calculus concepts to keep in mind are:

- Integration, symbolized by $\int$, is the math process of finding the area under a curve. A range will be specified on some integral symbols, and these define the points between which the area is taken.
- For sine waves, the action of integration causes the phase of the sine wave to shift 90°.

The Fourier series for any waveform can be expressed in the form:

$$f(t) = \frac{a_o}{2} + \sum_{n=1}^{\infty} [a_n \cos(n\omega t) + b_n \sin(n\omega t)] \tag{10-4}$$

Where: $a_o/2$ is the dc offset component (if any)
a_n and b_n represent the amplitudes of the harmonics (see below)
n is an integer (i.e., 0, 1, 2, 3, . . .)
$f(t)$ means that the series is "a function of time, t."

The sine and cosine wave amplitude coefficients in Eq. 10-4 (a_n and b_n) are expressed by:

$$a_n = \frac{2}{T} \int_0^t f(t) \cos(n\omega t) \, dt \tag{10-5}$$

and,

$$b_n = \frac{2}{T} \int_0^t f(t) \sin(n\omega t) \, dt \tag{10-6}$$

The amplitude terms are nonzero at the specific frequencies determined by the Fourier series. Because only certain frequencies, determined by integer n, are allowable, the spectrum of the periodic signal is said to be discrete.

The term $a_o/2$ in the Fourier series expression is the average value of $f(t)$ over one complete cycle (one period) of the waveform. In practical circuit terms, it is also the dc component of the waveform. When the waveform possesses halfwave symmetry (i.e., the peak amplitude above zero is equal to the peak amplitude below zero at every point in t, or $+V_m = |-V_m|$), there is no dc component, so $a_o/2 = 0$.

An alternative Fourier series expression replaces the $a_n\cos(n\omega t) + b_n\sin(n\omega t)$ with an equivalent expression of another form:

$$f(t) = \frac{2}{T} \sum_{n=1}^{\infty} C_n[n\omega t - \phi_n] \qquad \textbf{(10-7)}$$

where:

$$C_n = \sqrt{(a_n)^2 + (b_n)^2} \qquad \textbf{(10-8)}$$

$$\phi_n = \arctan\left(\frac{a_n}{b_n}\right) \qquad \textbf{(10-9)}$$

All other terms are as previously defined.

You can infer certain things about the harmonic content of a waveform by examining its symmetries. You would conclude from the above equations that the harmonics extend to infinity on all waveforms. Clearly, in practical systems a much less than infinite bandwidth is found, so some of those harmonics will be removed by the normal action of the electronic circuits. Also, it is sometimes found that higher harmonics might not be truly significant, so can be ignored. As n becomes larger, the amplitude coefficients a_n and b_n tend to become smaller. At some point, the amplitude coefficients are reduced sufficiently that their contribution to the shape of the wave is either negligible for the practical purpose at hand or is totally unobservable in practical terms. The value of n at which this occurs depends partially on the rise time of the waveform. *Rise time* is usually defined as the time required for the waveform to rise from 10 percent to 90 percent of its final amplitude.

The square wave represents another case altogether because it has a very fast rise time. Theoretically, the square wave contains an infinite number of harmonics, but not all of the possible harmonics are present. For example, in the case of the square wave only the odd harmonics are typically found (e.g., 3, 5, 7). According to some standards, accurately reproducing the square wave requires 100 harmonics, while others claim that 1000 harmonics are needed. Which standard to use may depend on the specifics of the application and the exact rise time needed.

Another factor that determines the profile of the Fourier series of a specific waveform is whether the function is odd or even. The *even* function is one in which $f(t) = f(-t)$, while for the odd function $-f(t) = f(-t)$. In the even function only cosine harmonics are present, so the sine amplitude coefficient b_n is zero. Similarly, in the *odd* function, only sine harmonics are present, so the cosine amplitude coefficient a_n is zero.

Waveform symmetry

Both symmetry and asymmetry can occur in several ways in a waveform, and those factors can affect the nature of the Fourier series of the waveform. In Fig. 10-5 we see the case of a waveform with a dc component. Otherwise, in terms of the Fourier series equation, the term a_o is nonzero. The dc component represents a case of asymmetry in a signal because the upper half (above the 0 V baseline) is not an ex-

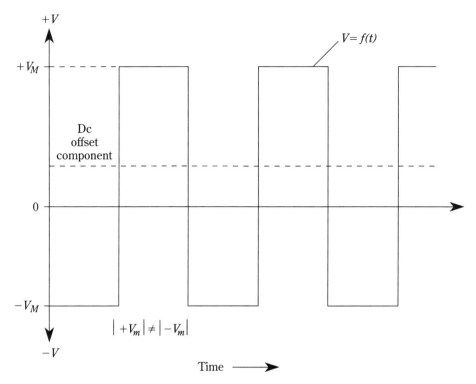

10-5 Square wave with dc offset.

act mirror image of the lower half. The amplitude of the pure square wave is V_a, and $|+V_a| = |-V_a|$. This dc offset can seriously affect instrumentation electronic circuits that are dc coupled, and thereby result in serious error in the output circuit.

Two different forms of symmetry are shown in Fig. 10-6. *Zero-axis symmetry* occurs when, on a point-for-point basis, the waveshape and amplitude above the zero baseline is equal to the amplitude below the baseline (or $|+V_m| = |-V_m|$). Both square wave (Fig. 10-6A) and triangle wave (Fig. 10-6B) are shown. When a waveform possesses zero-axis symmetry it will usually not contain even harmonics; only odd harmonics are present. This situation is found in square waves and triangle waves—for example, the Fourier spectrum for a square wave in Fig. 10-7.

Note in Fig. 10-6 that the positive and negative amplitudes are equal. This fact, and the shape, makes the waveform zero-baseline symmetrical. This is also called *half-wave symmetry*. In this type of symmetry the shape of the wave above the zero baseline is a mirror image of the shape of the waveform below the baseline. Half-wave symmetry also implies a lack of even harmonics.

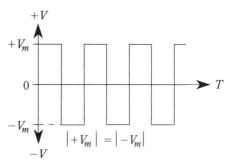

10-6 A. Square wave. B. Triangle wave.

A

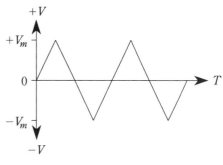

B

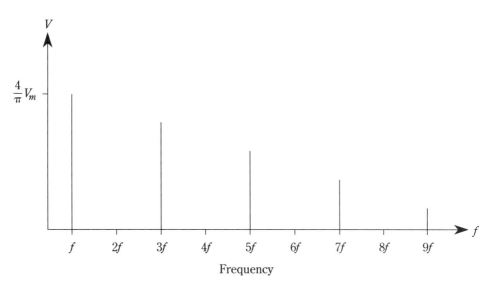

Frequency

10-7 Spectrum showing only odd harmonics.

An exception to the "no even harmonics" general rule is that there will be even harmonics present in the zero-axis symmetrical waveform if the even harmonics are in phase with the fundamental sine wave. This condition will neither produce a dc component nor disturb the zero-axis symmetry.

Quarter-wave symmetry exists when the left half and right half sides of the waveforms are mirror images of each other on the same side of the zero axis. The ideal square wave meets this requirement (Fig. 10-8A). The vertical dotted line divides the two halves on either side of the zero-baseline. Notice that in these cases the left and right halves are mirror images of each other.

Note in Fig. 10-8B, however, that waveforms that are not half-wave symmetrical can be quarter-wave symmetrical. In this example, the portion above the zero axis is like a square wave, and indeed the left- and right-hand sides are mirror images of each other. Similarly, below the zero axis the rounded waveform has a mirror-image relationship between left and right sides. In this case, there is a full set of even harmonics, and any odd harmonics that are present are in phase with the fundamental sine wave.

In the ideal, symmetrical square wave, the Fourier spectrum consists of the fundamental frequency (f) plus the odd-order harmonics ($3f$, $5f$, $7f$, etc). Furthermore, the harmonics are in phase with the fundamental. Theoretically, an infinite number of odd-number harmonics are present in the ideal square wave. However, in practi-

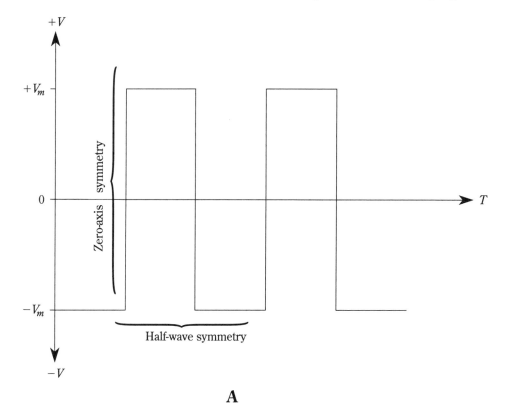

A

10-8A   Zero-axis and half-wave symmetry.

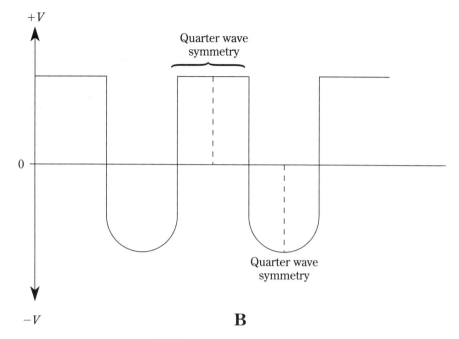

10-8B Quarter-wave symmetry.

cal square waves the "ideal" is considered satisfied with harmonics to about 999*f*. That ideal is almost never reached, however, due to the normal bandwidth limitations of the circuit. An indicator of harmonic content is the rise time of the square waves; the faster the rise time, the higher the number of harmonics.

The material in this chapter is only peripherally related to the overall topic of the book, but now you should have some good idea of how RC and RLC circuits can affect nonsinusoidal waveforms.

Experiment 10-1

Connect the circuit of Fig. 10-9. View square waves, triangle waves, sawtooth waves, and sine waves, at 10, 100, 1000, and 10,000 Hz. If your signal generator only has a square-wave output, then use it and ignore the other waveforms. Next, reverse the *R* and *C* elements and do the experiment again.

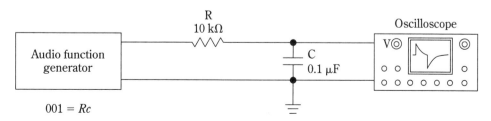

10-9 Test circuit.

11
Monostable and astable multivibrator circuits

IN THIS CHAPTER, WE WILL LOOK AT TWO DIFFERENT TYPES OF CIRCUITS THAT seem completely different, but are—in terms of circuitry—closely related. The *mono-stable multivibrator* circuit (also called the *one-shot* circuit) produces a single square output pulse for each input trigger pulse. The *astable multivibrator*, on the other hand, produces a chain of square output pulses. Both of these circuits find extensive application in electronic instruments and circuits. Additional information on this class of circuits, but that are based on a special integrated circuit (the 555 timer), is found in chapter 17.

Monostable multivibrator circuits

The *monostable multivibrator* (MMV), or one-shot, has two permissible output states (high and low), but only one of them is stable. The MMV produces one output pulse in response to an input trigger signal (Fig. 11-1); in this case, a negative-going trigger pulse (V_t) results in a positive-going output pulse (V_o). The output pulse (V_o) has a duration, T, in which the output is in the quasi-stable state. The MMV is also known under several alternate names: *one-shot, pulse generator,* and *pulse stretcher*. The latter name derives from the fact that the output duration T is longer than the trigger pulse ($T > T_t$) as in Fig. 11-1.

Monostable multivibrators find a wide variety of applications in electronic circuits. Besides the pulse stretcher mentioned above, the MMV also serves to lock out unwanted pulses. Figure 11-2 shows that the output responds to only the first trigger pulse. The next two pulses occur during the active time, T, so are ignored. Such an MMV is said to be nonretriggerable. A common application of this feature is in switch-contact "debouncing" in keyboards. All mechanical switch contacts bounce a few times on closure, creating a short run of exponentially decaying pulses. If an

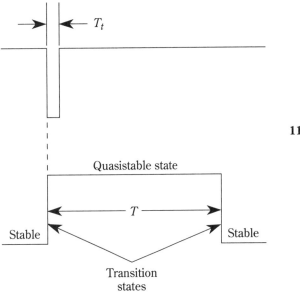

11-1 Monostable multivibrator (one-shot) triggering operation.

MMV is triggered by the first pulse from the switch, and if the MMV remains quasi-active long enough for the bouncing pulses to die out, then the MMV output signal becomes the debounced "switch closure." The main requirement is that the MMV duration be longer than the switch-contact bounce pulse train (5 ms is generally considered adequate for most switch types).

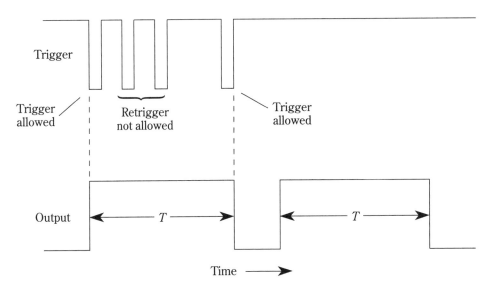

11-2 Effect of additional trigger pulses on nonretriggerable circuit.

The range of possible MMV applications is too broad for detailed discussion here, so only a general set of categories can be presented. These application categories include: pulse generation, pulse stretching, contact debouncing, pulse signal cleanup, switching, and synchronization of circuit functions (especially digital).

Figure 11-3A shows the circuit for a nonretriggerable monostable multivibrator based on the common operational amplifier, while the timing diagram is shown in Fig. 11-3B. This circuit is based on the voltage-comparator circuit. When there is no feedback, the effective voltage gain of an op amp is its open-loop gain (A_{vol}). When both −IN and +IN are at the same potential, the differential input voltage (V_{id}) is zero, so the output is also zero. But if $V_{(-IN)}$ does not equal $V_{(+IN)}$, the high gain of the amplifier forces the output to either its positive or negative saturation values. If $V_{(-IN)} > V_{(+IN)}$, the op amp sees a positive differential input signal, so the output saturates at $−V_{sat}$. However, if $V_{(-IN)} < V_{(+IN)}$, the amplifier sees a negative differential input signal and the output saturates to $+V_{sat}$. The operation of the MMV depends on the relationship of $V_{(-IN)}$ and $V_{(+IN)}$.

Four states of the monostable multivibrator must be considered: stable state, transition state, quasistable state, and refractory state.

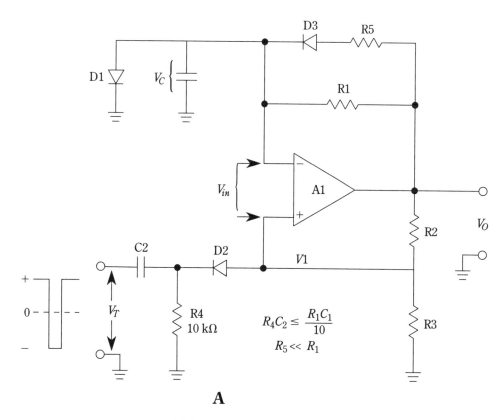

A

11-3A Op-amp monostable circuit.

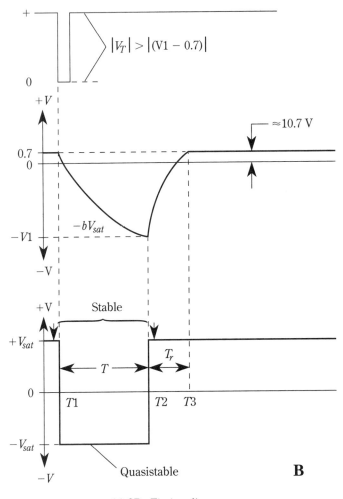

11-3B Timing diagram.

Stable state

The output voltage V_o is initially at $+V_{sat}$. Capacitor C_1 will attempt to charge in the positive-going direction because $+V_{sat}$ is applied to the R_1C_1 network. However, because of diode D1 shunted across C_1, the voltage across C_1 is clamped to $+V_{D1}$. For a silicon diode such as the 1N914 or 1N4148 $+V_{D1}$ is about +0.7 Vdc. Thus, the inverting input (−IN) is held to +0.7 Vdc during the stable state. The noninverting input (+IN) is biased to a level V_1, which is:

$$V_1 = \frac{R_3(+V_{sat})}{R_2 + R_3} \qquad (11\text{-}1)$$

or, in the special case of $R_2 = R_3$:

$$V_1 = \frac{+V_{sat}}{2} \tag{11-2}$$

The factor $R_3/(R_2 + R_3)$ is often designated by the Greek letter beta (β), so:

$$\beta = \frac{R_3}{R_2 + R_3} \tag{11-3}$$

Therefore:

$$V_1 = \beta(+V_{sat}) \tag{11-4}$$

The amplifier (A1) sees a differential input voltage (V_{id}) of ($V_1 - V_{D1}$), or ($V_1 -$ 0.7) V. Using the previous notation:

$$V_{id} = \frac{R_3(+V_{sat})}{R_2 + R_3} - 0.7 \tag{11-5}$$

As long as $V_1 > V_{D1}$, the amplifier effectively sees a negative dc differential voltage at the inverting input, so (with its high open-loop gain, A_{vol}) will remain saturated at $+V_{sat}$. For purposes of this discussion the amplifier is a type 741 operated at dc power-supply potentials of ± 12 Vdc, so V_{sat} typically will be ± 10 V.

Transition state

The input trigger signal (V_t) is applied to the MMV of Fig. 11-3A through RC network R_4C_2. The general design rule for this network is that its time-constant should be not more than one-tenth the time-constant of the timing network:

$$R_4C_2 < \frac{R_1C_1}{10} \tag{11-6}$$

At time T1 (see Fig. 11-3B), trigger signal V_t makes an abrupt high to low transition to a peak value that is less than ($V_1 - 0.7$) V. Under this condition the polarity of V_{id} is now reversed and the inverting input now sees a positive voltage: ($V_1 + V_t -$ 0.7) is less than V_{D1}. The output voltage V_o now snaps rapidly to $-V_{sat}$. The fall time of the output signal is dependent upon the slew rate and the open-loop gain of the operational amplifier, A1.

Quasistable state

The "output signal" from the MMV is the quasistable state between T_1 and T_2 in Fig. 11-3B. It is called "quasistable" because it does not change over $T = T_2 - T_1$, but when T expires the MMV "times out" and V_o reverts to the stable state ($+V_{sat}$).

During the quasistable time, D1 is reverse biased, and capacitor C_1 discharges from $+0.7$ Vdc to zero and then recharges towards $-V_{sat}$. When $-V_o$ reaches $-V_1$, however, the value of V_{id} crosses zero and that change forces V_o to snap once again to $+V_{sat}$.

Appealing to Eq. 11-6 makes it possible to derive the timing equation for the MMV. The timing capacitor must charge from an initial value (V_{C1}) to a final value (V_{C2}) in time T. The question is: "What value of R_1C_1 will cause the required transitions?" Consider the case $R_2 = R_3$ $(V_1 = 0.5\ V_{sat})$:

$$R_1C_1 = \frac{-T}{\ln\left(\dfrac{V_{sat} - V_{c2}}{V_{sat} - V_{c1}}\right)} \tag{11-7}$$

$$R_1C_1 = \frac{-T}{\ln\left(\dfrac{V_{sat} - ((0.5)\ (V_{sat} + 0.7))}{V_{sat} - 0.7}\right)} \tag{11-8}$$

and, for the case where $V_{sat} = 10$ Vdc:

$$R_1C_1 = \frac{-T}{\ln\left(\dfrac{10\ \text{Vdc} - ((0.5)\ (10\ \text{Vdc} + 0.7))}{10\ \text{Vdc} - 0.7}\right)} \tag{11-9}$$

$$= \frac{-T}{\ln\left(\dfrac{10\ \text{Vdc} - 5.35\ \text{V}}{10\ \text{Vdc} - 0.7\ \text{V}}\right)}$$

$$= \frac{-T}{\ln\left(\dfrac{4.65}{9.3}\right)}$$

$$= \frac{-T}{\ln(0.5)}$$

$$= \frac{-T}{-0.69}$$

Thus,

$$T = 0.69\ R_1C_1 \tag{11-10}$$

Equation 11-10 represents the special case in which $B = \frac{1}{2}$ (i.e., $R_2 = R_3$). Although $R_2 = R_3$ may be the usual case for this class of circuit, R_2 and R_3 might not be equal in other cases. A more generalized expression is:

$$RC = \frac{T}{\ln\left(\dfrac{1 + 0.7V/V_{sat}}{1 - \beta}\right)} \qquad \textbf{(11-11)}$$

In which:

$$\beta = \frac{R_3}{R_2 + R_3} \qquad \textbf{(11-12)}$$

When the quasistable state times out, the circuit status returns to the stable state (where it remains dormant until triggered again).

Design example 11-1

A monostable multivibrator is constructed such that $R_2 = 10\ \text{k}\Omega$, and $R_3 = 4.7\ \text{k}\Omega$. If the active device (A1) is a 741 op amp, and the dc power supplies are ±12 Vdc, then $\pm V_{sat} = \pm10$ V. Find the RC time constant to produce a 5-s output pulse. Also, propose values for sample components.

    1. First, calculate β:

$$\beta = \frac{R_3}{R_2 + R_3}$$

$$= \frac{4.7\ \text{k}\Omega}{(10\ \text{k}\Omega + 4.7\ \text{k}\Omega)}$$

$$= \frac{4.7}{14.7}$$

$$= 0.32$$

    2. Calculate $R_1 C_1$:

$$RC = \frac{T}{\ln\left(\dfrac{1 + 0.7\ V/V_{sat}}{1 - \beta}\right)}$$

$$R_1 C_1 = \frac{5\ \text{s}}{\ln\left(\dfrac{1 + (0.7\ V/10\ V)}{1 - 0.32}\right)}$$

$$= \frac{5\ \text{s}}{\ln\left(\dfrac{1.07}{0.68}\right)}$$

$$= \frac{5 \text{ s}}{\ln (1.57)}$$

$$= \frac{5 \text{ s}}{0.453}$$

$$= 11.04 \text{ s}$$

3. The time constant of $R_1 C_1$ is 11.04 s. Select a standard-value capacitor and find the resistor values that will give the correct value of $R_1 C_1$.

Candidates: C_1 = 10 μF, R_1 = 1.1 MΩ (1 MΩ)
C_1 = 6.8 μF, R_1 = 1.62 MΩ (1.5 MΩ)
C_1 = 4.7 μF, R_1 = 2.35 MΩ (2.2 MΩ)

If the accuracy of the pulse is not critical, then select one of the standard-value capacitors and the nearest standard-value resistor (shown above in parentheseis).

Experiment 11-1

This experiment explores the monostable multivibrator (one-shot) circuit. You will not need an oscilloscope for this experiment, but one can be used to monitor either V_c or V_o if you want.

1. Construct the circuit of Fig. 11-EXP1 using the values shown for R1C1 (alternate values from the Design Example above can be subsituted if you wish). The output indicator consists of two light-emitting diodes (LEDs) with their associated current-limiting resistors (R_6 and R_7). A 741 operational amplifier may be used for A1. Switch S1 and pull-up resistor act as a source for trigger signals.
2. Turn the power on. Only one LED should be lighted; this is the stable state.
3. Press S1 momentarily, and then release. The output should snap to the other output state, causing the lighted LED to extinquish, and the unlighted LED to light up. This is the quasistable state.
4. The second LED should remain lighted for about 5 s. Time the actual quasistable output duration with a sweep second hand.
5. After the one-shot "times out," i.e., returns to the stable state, then trigger the circuit again (press S1); release S1 and then immediately press it again. If you time the circuit, you will find that the output duration does not increase following the second closing of S1.

Refractory period

At time t_2 the output signal voltage V_o switches from $-V_{sat}$ to $+V_{sat}$. Although the output has timed out, the MMV is not yet ready to accept another trigger pulse. The *refractory state* between t_2 and t_3 is characterized by the output being in the stable state, but the input is unable to accept a new trigger input stimulus. The refractory period must await the discharge of C_1 under the influence of the output voltage to satisfy $V_1 < (V_1 - 0.7)$ V.

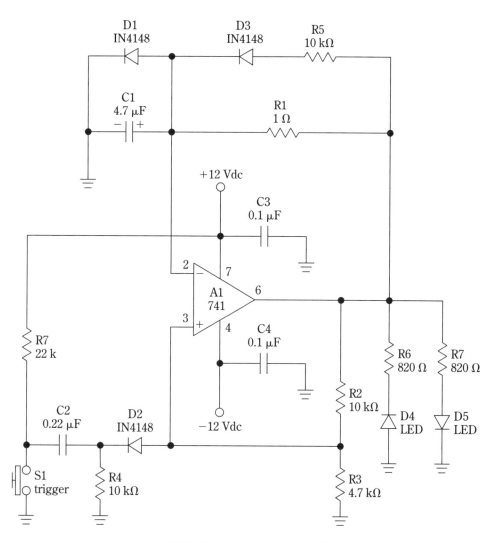

11-EXP1 Circuit for experiment 11-1.

Astable multivibrators

The square-wave oscillator is, perhaps, the most common form of waveform generator other than the sine-wave oscillator. These circuits are easily designed from resistor-capacitor networks and a variety of active devices (we will use operational
amplifiers). In this chapter, you will learn how those circuits work and how to design
them, and you will perform some experiments based on these circuits.

Astable (free-running) circuits

An *astable multivibrator* (AMV) is free-running; i.e., it is a self-starting circuit which outputs a waveform that repeats itself without being either triggered or re-triggered. The output of the AMV is a periodic (i.e., repeats itself regularly) pulse or wave train. In a periodic signal the wave repeats itself indefinitely until the circuit is either turned off or otherwise inhibited.

Astable multivibrators are oscillators. Waveforms available from the AMV include square waves, triangle waves, and sawtooth waves. Sine waves are also available from oscillator circuits, but most of those circuits operate differently from the others.

Nonsinusoidal waveform generators

Nonsinusoidal AMV circuits will produce square, triangle, or sawtooth waves, depending on the design. When combined with a monostable multivibrator (MMV), in order to produce variable-width pulses (of duration less than the period of the square wave), a variable pulse generator circuit results. Because the square-wave generator is the most basic form, the discussion of AMV circuits begins with square waves—the subject of this chapter.

Square waves

Figure 11-4 shows the classic square wave. Each time interval of the wave is quasistable, so you might conclude that the square-wave generator has no stable states (hence it is astable). The waveform snaps back and forth between $-V$ and $+V$, dwelling on each level of a duration of time (t_a or t_b). The period, T, is the sum of these dwell times:

$$T = t_a + t_b \qquad\qquad\qquad \textbf{(11-13)}$$

Where: T is the period of the square wave (t_1 to t_3)
  t_a is the interval t_1 to t_2
  t_b is the interval t_2 to t_3

The *frequency of oscillation* (F) is the reciprocal of T:

$$F = \frac{1}{T} \qquad\qquad\qquad \textbf{(11-14)}$$

Where: F is in hertz (Hz) and T is in seconds (s).

The ideal square wave is both base-line and time-line symmetrical. That means that $|+V| = |-V|$ and $t_a = t_b$. Under time-line symmetry $t_a = t_b = t$, so $T = 2t$ and $f = \frac{1}{2}t$.

The circuit for an operational amplifier square wave generator is shown in Fig. 11-5A. The operation of this circuit depends upon the relationship between $V_{(-in)}$ and $V_{(+in)}$. In the circuit of Fig. 11-5A the voltage applied to the noninverting input ($V_{(+in)}$) is determined by a resistor voltage divider, R_2 and R_3. This voltage is called V_1 in Fig. 11-5A and is:

$$V_1 = \frac{V_o R_3}{R_2 + R_3} \qquad\qquad\qquad \textbf{(11-15)}$$

or, when V_o is saturated,

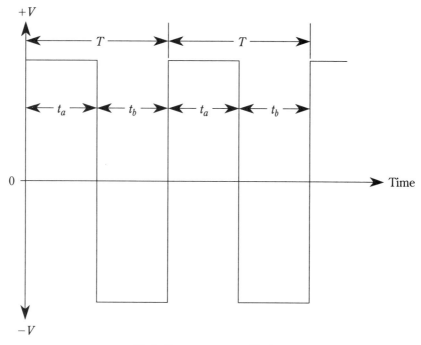

11-4 Square-wave oscillation.

$$V_1 = \frac{V_{sat}R_3}{R_2 + R_3} \tag{11-16}$$

Once again, the factor R3/(R2 + R3) is often designated B:

$$\beta = \frac{R_3}{R_2 + R_3} \tag{11-17}$$

Because Eq. 11-17 is always a fraction, $V_1 < V_{sat}$ and V_1 is of the same polarity as V_{sat}.

The voltage applied to the inverting input $(V_{(-in)})$ is the voltage across capacitor C_1, i.e., V_{C1}. This voltage is created when C_1 charges under the influence of current I, which in turn is a function of V_o and the time constant of R_1C_1. Timing operation of the circuit is shown in Fig. 11-5B.

At turn-on $V_{C1} = 0$ V and $V_o = +V_{sat}$, so $V_1 = +V_1 = \mathrm{B}(+V_{sat})$. Because $V_{C1} < V_1$, the op amp sees a negative differential input voltage so the output remains at $+V_{sat}$. During this time, however, V_{C1} is charging towards $+V_{sat}$ at a rate of:

$$V_{C1} = V_{sat}(1 - \epsilon^{-t2/R_1C_1}) \tag{11-18}$$

When V_{C1} reaches $+V_1$, however, the op amp sees $V_{C1} = V_1$, so $V_{id} = 0$. The output now snaps from $+V_{sat}$ to $-V_{sat}$ (time t_2 in Fig. 11-5B). The capacitor now begins to discharge from $+V_1$ toward zero, and then recharges towards $-V_{sat}$. When it reaches $-V_1$, the inputs are once again zero, so the output again snaps to $+V_{sat}$. The output continuously snaps back and forth between $-V_{sat}$ and $+V_{sat}$, thereby producing a square-wave output signal.

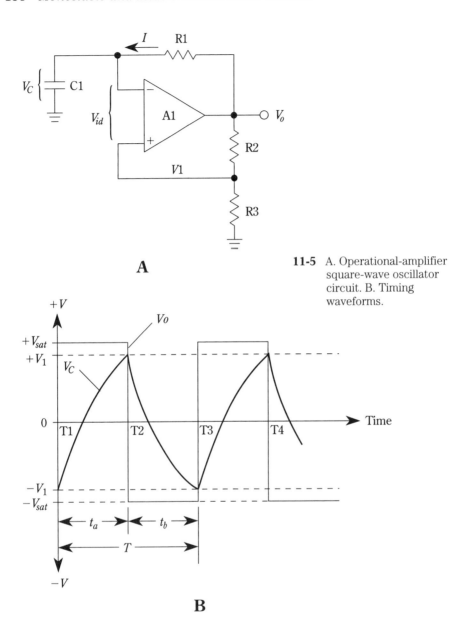

A

11-5 A. Operational-amplifier square-wave oscillator circuit. B. Timing waveforms.

B

The time constant required to charge from an initial voltage V_{C1} to an end voltage V_{C2} in time t is defined by:

$$RC = \frac{-T}{\ln\left(\dfrac{V - V_{C2}}{V - V_{C1}}\right)} \tag{11-19}$$

In Fig. 11-5A the RC time constant is R_1C_1. From Fig. 11-5B it is apparent that, for interval t_a, $V_{C1} = -\beta V_{sat}$, $V_{C2} = +\beta V_{sat}$ and $V = V_{sat}$. To calculate the period T, then:

$$2R_1C_1 = \cfrac{-T}{\ln\left(\cfrac{V_{sat} - \beta V_{sat}}{V_{sat} - (-\beta V_{sat})}\right)} \tag{11-20}$$

or, rearranging Eq. 11-20:

$$-T = 2R_1C_1\ln\left(\frac{V_{sat} - \beta V_{sat}}{V_{sat} - (-\beta V_{sat})}\right) \tag{11-21}$$

$$-T = 2R_1C_1\ln\left(\frac{1 - \beta}{1 + \beta}\right) \tag{11-22}$$

$$T = 2R_1C_1\ln\left(\frac{1 + \beta}{1 - \beta}\right) \tag{11-23}$$

Because $\beta = R_3/(R_2+R_3)$:

$$T = 2R_1C_1\ln\left(\cfrac{1 + \left(\cfrac{R_3}{R_2 + R_3}\right)}{1 - \left(\cfrac{R_3}{R_2 + R_3}\right)}\right) \tag{11-24}$$

which reduces to:

$$T = 2R_1C_1\ln\left(\frac{2R_2}{R_3}\right) \tag{11-25}$$

Equation 11-25 defines the frequency of oscillation for any combination of R_1, R_2, R_3, and C_1. In the special case $R_2 = R_3$, $\beta = 0.5$, so:

$$T = 2R_1C_1\ln\left(\frac{1 + 0.5}{1 - 0.5}\right)$$

$$= 2\,R_1C_1\ln\left(\frac{1.5}{0.5}\right)$$

$$= 2R_1C_1\ln(3)$$
$$= 2R_1C_1\,(1.1)$$
$$= 2.2R_1C_1$$

$$T = 2.2R_1C_1 \tag{11-26}$$

and, because $f = 1/T$:

$$f = \frac{0.4545}{R_1C_1} \tag{11-27}$$

Experiment 11-2   Basic square-wave generator circuit

This experiment looks at the basic form of square-wave generator circuit. It will produce a symmetrical square wave in which the amplitudes of positive and negative
swings are equal to each other. This experiment does not require an oscilloscope,
because it uses a pair of light-emitting diodes (LEDs) as output indicators (see the
circuit in Fig. 11-EXP2).

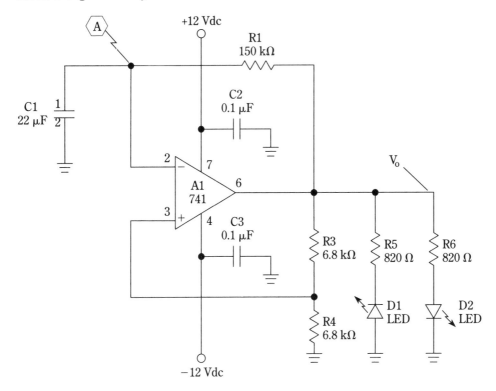

11-EXP2   Circuit for experiment 11-2.

An LED is a special pn junction diode that emits light when it is forward biased.
In this experiment, we will use two red LEDs wired back to back. In Fig. 11-EXP2,
the two LEDs are shown connected across the output of the operational amplifier.
Each LED has a resistor (R_5 and R_6) in series to limit the current to less than 20 mA.
When the output voltage V_o is low (i.e., at $-V_{sat}$), then diode D1 is forward biased,
and this diode emits light; when the output is high (i.e., at $+V_{sat}$), then the other
diode (D2) is forward biased (turned on).

   1. Connect the circuit of Fig. 11-EXP2. Use a capacitor for C_1 between 16 μF
      and 30 μF, and a resistance of 120 kΩ to 220 kΩ.

2. Calculate the time $T = 2.2R_1C_1$.
3. Turn on the circuit by applying ± 12 Vdc.
4. If the circuit is working properly, then the LEDs will alternate on and off at a rate equal to time T; measure the time with a watch second hand. If only one LED turns on, or if neither LED turns on, then assume either a wiring error or a bad op amp.
5. Shunt a second timing resistor across R_1 such that its value is approximately one-half the first value; i.e., parallel another resistor of the same value across R_1.
6. Calculate the new time T.
7. Measure T with a sweep second hand, and compare to the predicted value.
8. Change the values of the capacitor and resistor to various combinations and repeat the experiment. If you select an RC time constant that is too short, then the LEDs will blink on and off too rapidly.

Experiment 11-3

This experiment uses the same circuit as EXP. 11-2, but uses an oscilloscope to examine the output waveform. If you have a two-channel oscilloscope, then connect one probe across the output of the op amp, and the other across the timing capacitor.

1. Build the circuit of Fig. 11-EXP3 using the following values for R_1 and C_1:
$$R_1 = 47 \text{ k}\Omega$$
$$C_1 = 0.01 \text{ }\mu\text{F}$$
2. Calculate the value of T.
3. Examine the output waveform on an oscilloscope.
4. Measure the period of the waveform, and compare with step 2.
5. If you have a two-channel oscilloscope, then connect the second input across the timing capacitor, C_1.
6. Compare the capacitor waveform with the output waveform, paying attention to the amplitude and timing aspects.
7. What can you tell about the feedback voltage from this comparison? Does this tell you the ratio of R_3/R_4?

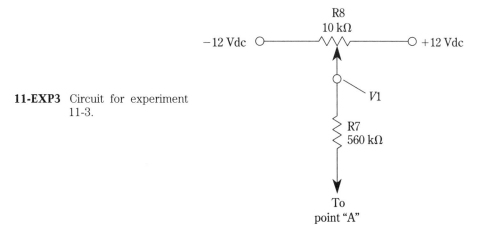

11-EXP3 Circuit for experiment 11-3.

Time-asymmetrical square waves

The circuit of Fig. 11-5A produces time line symmetrical square waves (i.e., $t_a = t_b$). If time-line asymmetrical square waves are required, then a circuit such as either Fig. 11-6 or 11-8A is required. The circuit in Fig. 11-6 uses a potentiometer (R_4) and a fixed resistor (R5) to establish a variable duty cycle asymmetry. The circuit is similar to Fig. 11-5A, but with an offset circuit (R_4/R_5) added. The assumptions are $R_5 = R_1$, and $R_4 \ll R_1$. If V_a is the potentiometer output voltage, C_1 charges at a rate of $(R_1/2)C_1$ towards a potential of ($V_a + V_{sat}$). After output transition, however, the capacitor discharges at the same $(R_1/2)C_1$ rate towards ($V_a - V_{sat}$). The two interval times are therefore different; t_a and t_b are no longer equal.

Figure 11-7 shows three extremes of V_a: $V_a = +V$ (Fig. 11-7A), $V_a = 0$ (Fig. 11-7B) and $V_a = -V$ (Fig. 11-7C). These traces represent very long, equal, and very short duty cycles, respectively.

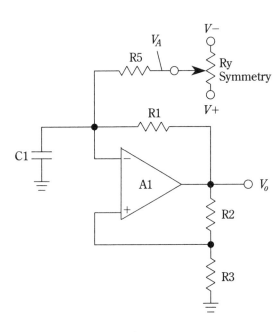

11-6 Duty cycle variation using an offset voltage.

Experiment 11-4   Time-asymmetrical square wave

This circuit examines the asymmetrical square-wave generator circuit. The symmetry of the waveform is set by a potentiometer and fixed resistor added to the circuit of Fig. 11-EXP2. These components (Fig. 11-EXP3) are connected into the original circuit at point "A." All other connections remain the same.

1. Build and test the circuit used in Exp. 11-2 (i.e., Fig. 1-EXP2); use the values recommended and record the timing. Turn the circuit off when testing is finished.
2. Add the circuit of Fig. 11-EXP3.
3. Turn the circuit on.

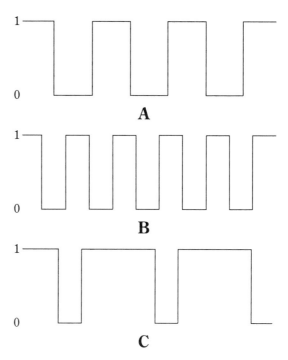

11-7 Duty cycle variation at three settings of the potentiometer: A. Maximum negative. B. Centered. C. Maximum positive.

4. Use a voltmeter to measure the voltage at the wiper of the potentiometer (V_1 in Fig. 11-EXP3). Set the voltage $V_1 = 0$.
5. Observe the output LEDs, and compare the timing with the original configuration.
6. Adjust the potentiometer R_8 to set V_1 to its maximum positive value.
7. Repeat step 5.
8. Readjust the potentiometer for maximum negative voltage at V_1.
9. Repeat step 5.
10. Compare the timing data for all three conditions.
11. Experiment with various settings of R_8, and with different values of R_7.

If you own an oscilloscope, you can perform this experiment using the component values used in experiment 11-3.

The circuit of Fig. 11-8A also produces asymmetrical square waves, but the duty cycle is fixed instead of variable. Once again the basic circuit is like Fig. 11-5A, but with added components. In Fig. 11-8A, the RC timing network is altered such that the resistors are different on each swing of the output signal. During t_a, $V_a = +V_{sat}$, so diode D1 is forward biased and D2 is reverse biased. For this interval:

$$t_a = (R_{1A})\,(C_1)\,\ln\!\left(1 + \frac{2R_2}{R_3}\right) \qquad (11\text{-}28)$$

During the alternate half-cycle (t_b), the output voltage V_o is at $-V_{sat}$, so D1 is reverse biased and D2 is forward biased. During this interval R_{1B} is the timing resistor, while R_{1A} is effectively out of the circuit. The timing equation is:

$$t_b = (R_{1B})\,(C_1)\,\ln\!\left(1 + \left(\frac{2R_2}{R_3}\right)\right) \tag{11-29}$$

The total period, T, is $t_a + t_b$, so:

$$T = (R_{1A})\,(C_1)\,\ln\!\left(1 + \left(\frac{2R_2}{R_3}\right)\right) + (R_{1B})\,(C_1)\,\ln\!\left(1 + \left(\frac{2R_2}{R_3}\right)\right) \tag{11-30}$$

Collecting terms:

$$T = (R_{1A} + R_{1B})\,C_1\,\ln\!\left(1 + \frac{2R_2}{R_3}\right) \tag{11-31}$$

Equation 11-31 defines the oscillation frequency of the circuit in Fig. 11-8A. Figures 11-8B and 11-8C show the effects of two values of R_{1A}/R_{1B} ratio. In Fig. 11-8B the ratio $R_{1A}/R_{1B} = 3{:}1$, while in Fig. 11-8C the ratio $R_{1A}/R_{1B} = 10{:}1$.

The effect of this circuit on capacitor charging can be seen in Fig. 11-9. A relatively low R_{1A}/R_{1B} ratio is seen in Fig. 11-9A. Notice in the lower trace that the capacitor charge time is long compared with the discharge time. The effect is seen even better for the case of a high R_{1A}/R_{1B} ratio (Fig. 11-9B).

Experiment 11-5

In this experiment we deal with the other method for producing asymmetrical waveforms: the diode-controlled switch. This experiment is based on the original experiment circuit shown in Fig. 11-EXP2.

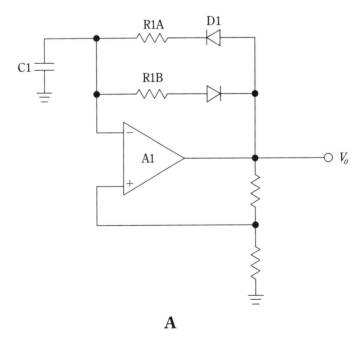

A

11-8A Fixed asymmetrical duty cycle circuit. B. 3:1 ratio. C. 10:1 ratio.

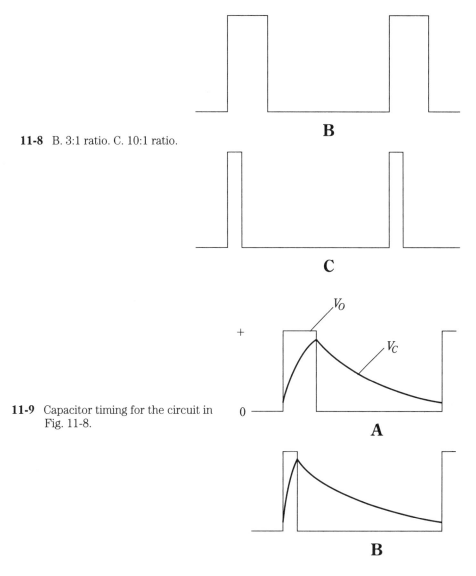

11-8 B. 3:1 ratio. C. 10:1 ratio.

11-9 Capacitor timing for the circuit in Fig. 11-8.

1. Wire and test the circuit of Fig. 11-EXP2. Turn the circuit off when testing is completed.
2. Modify the circuit of Fig. 11-EXP.2 in the manner shown in Fig. 11-EXP.4. First, place a diode in series with R1 (which is now relabeled "R1a"); next, add a second feedback circuit consisting of R1b and D2. Note that the diodes are connected backwards with respect to each other. The direction of the arrow with respect to the actual diode is shown in the inset to Fig. 11-EXP4.
3. Turn on the circuit.
4. Observe the timing of both LEDs and compare their respective on periods. Use a watch sweep second hand to make timing measurements.
5. Change the value of R_{1b} to both higher and lower values, and compare the results to the original measurement.

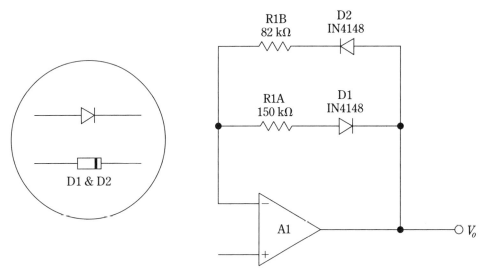

11-EXP4 Circuit for experiment 11-4.

If you own an oscilloscope, you might wish to perform this experiment using the component values used in experiment 11-2.

Output voltage limiting

The standard op-amp MMV or AMV circuit sometimes produces a relatively sloppy square output wave. By adding a pair of back-to-back zener diodes (Fig. 11-10) across the output, however, the signal can be cleaned up (although at the expense of amplitude). For each polarity the output signal sees one forward-biased and one reverse-biased zener diode. On the positive swing, the output voltage is clamped at $[V_{Z1} + 0.7]$ V. The 0.7-V factor represents the normal junction potential across the forward-biased diode (D2). On negative swings of the output signal, the situation reverses. The output signal is clamped to $[-(V_{Z2} + 0.7)]$ V.

Experiment 11-6

Perform Experiment 11-3, but modify the circuit as shown previously in Fig. 11-10. The series resistor (R4) should have a value of 1000 Ω, while the zener diodes should be 5.6- to 6.8-V units (make them identical). Compare the waveforms (both as to rise time, rounding, and amplitude) at the output of the op amp (pin 6) and across the diodes.

Square waves from sine waves

Figure 11-11 shows a method for converting sine waves to square waves. The circuit is shown in Fig. 11-11A, while the waveforms are shown in Fig. 11-11B. The circuit is an operational amplifier connected as a comparator. Because the op amp has no negative feedback path, the gain is very high (i.e., A_{vol}); in op amps, gains of 20,000

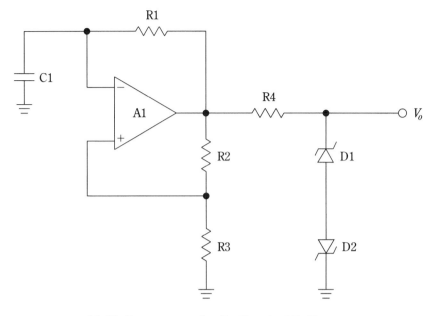

11-10 Square-wave circuit with output limiting.

to 2,000,000 (250,000 typical) are found. Thus, a voltage difference across the input terminals of only a few millivolts will saturate the output.

The input waveform is a sine wave. Because the noninverting input is grounded (Fig. 11-11A), the output of the op amp is zero only when the input signal voltage is also zero. When the sine wave is positive, the output signal will be at $-V_o$; when the sine wave is negative, the output signal will be at $+V_o$. The output signal will be a square wave at the sine wave frequency, with a peak-to-peak amplitude of $[(+V_o) - (-V_o)]$.

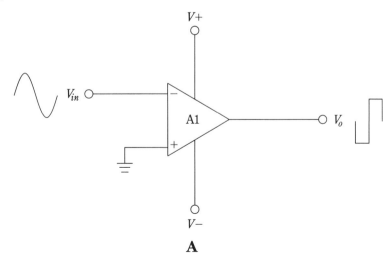

A

11-11A Converting sine waves and other waveforms into square waves: Basic circuit is a voltage comparator.

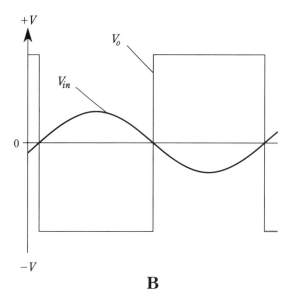

B

11-11B Input and output waveforms.

Project 11-1   Variable square-wave generator

This circuit puts into practice what you have learned about square-wave generator circuits. The circuit of Fig. 11-P1 is a two-stage oscillator-amplifier based on the RCA/GE CA-3240 BiMOS operational amplifier. The CA-3240 is a dual version of the famous CA-3140 device, and is similar in form (but not capability) to the lower grade 1456 devices (which are 741-family op amps).

The frequency of the square-wave generator circuit is set by a complex network consisting of R1, R2 (the adjust-frequency control), switch S1 (which selects decade differences in capacitance), and timing capacitors C1 through C6. Switch S1 selects the frequency range, while potentiometer R2 selects the exact frequency within that range. If a scale connected to R2 is calibrated for one range, then it will also serve for the other ranges, because there is a 10:1 ratio between ranges.

The output circuit consists of the output limiting circuit for A1a (zener diodes D1 and D2), a level-set control, and a unity-gain noninverting follower used as a buffer amplifier.

The ±12 Vdc power supplies are connected to the op amp at pins 4 (for −12 V) and 8 (for +12 V). Each power-supply terminal is bypassed by a 0.1-μF capacitor. This capacitor must be placed as close as possible to the body of the op amp. If the circuit tends to ring or oscillate at a high frequency (>100 kHz), then increase the value of the bypass capacitors.

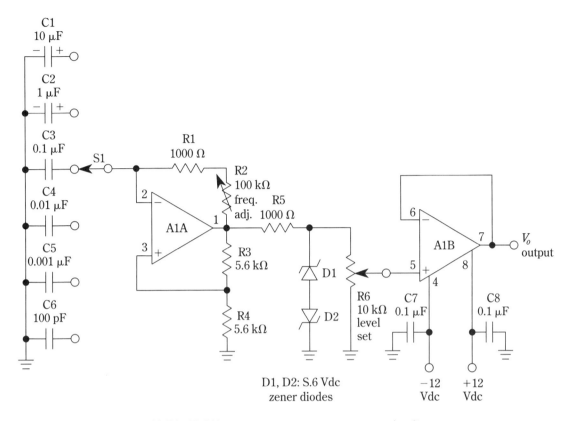

11-P1 Multifrequency square-wave generator circuit.

12

Electronic integrators
and differentiators

INTEGRATION AND DIFFERENTIATION ARE IMPORTANT MATHEMATICAL PROCESSES to electronic instrumentation and signal processing. These processes are inverses of each other, so a function that is first integrated and then differentiated returns to the original function. A similar relationship occurs when a function is differentiated and then integrated. Such is the normal nature of mathematically inverse processes.

These processes are seen elsewhere in electronics, but sometimes under different names. The differentiator is sometimes called a *rate-of-change circuit*, or, if the time constant is correct, a *high-pass filter*. Similarly, the integrator might be called a *time-averager* circuit or *low-pass filter*.

Let's consider an example of integration in electronic instruments. In Fig. 12-1A a voltage represents a pressure-transducer output, in this particular case the output of a human blood-pressure transducer. Such sensors are common in hospital intensive-care units. Notice that the pressure voltage varies with time from a low ("diastolic") to a high ("systolic") between T_1 and T_2 (which interval represents one complete cardiac cycle). If you want to know the mean arterial blood pressure (MAP), then you would want to find the area under the pressure-VS-time curve over one cardiac cycle (Fig. 12-1B).

An electronic integrator circuit serves to compute the time average of the analog voltage waveform that represents the time-varying arterial blood pressure. In an electronic blood-pressure monitoring instrument, a voltage serves to represent the pressure. If, for example, a scaling factor of 10 mV/mmHg is used (as is commonly the case in medical devices), then a pressure of 100 mmHg is represented by a potential of 1000 mV, or 1.000 V. This voltage will vary over the range 800 mV to 1200 mV for the case shown in Fig. 12-1C (pressure varies from 80 mmHg to 120 mmHg).

Electronic integrators and differentiators affect signals in different ways. Figure 12-2 shows the example of a square wave applied to the inputs of an integrator and differentiator. The integrator output is shown in Fig. 12-2, while the differentiator output is shown in Fig. 12-3(A, B).

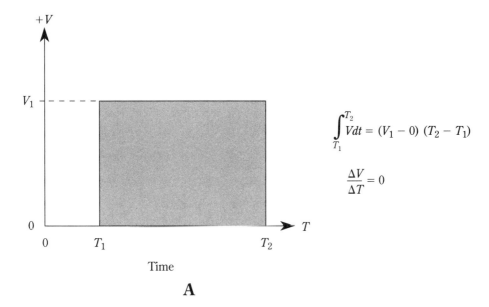

$$\int_{T_1}^{T_2} V dt = (V_1 - 0)\,(T_2 - T_1)$$

$$\frac{\Delta V}{\Delta T} = 0$$

A

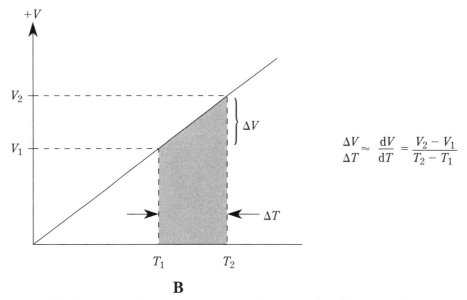

$$\frac{\Delta V}{\Delta T} \approx \frac{\mathrm{d}V}{\mathrm{d}T} = \frac{V_2 - V_1}{T_2 - T_1}$$

B

12-1 A. Area under a constant function. B. Area under a linear function.

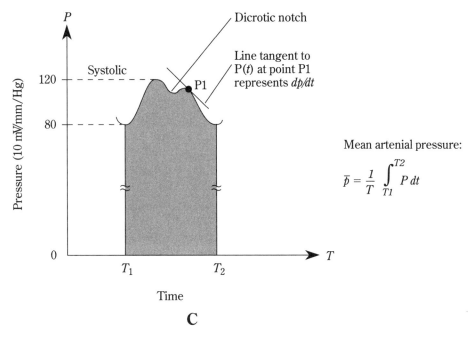

12-1C Area under a time-varying function.

First consider the operation of the integrator circuit. The integrator output waveform in Fig. 12-2 shows a constant positive-going slope between T_1 and T_2. The steepness of the slope is dependent upon the amplitude of the input square wave, but the line is linear. You can see from curve "B" in Fig. 12-2 that the square wave into the integrator produces a triangle waveform.

Now consider the operation of the differentiator circuit (see output waveform in Fig. 12-3). At time T_1 the square wave makes a positive-going transition to maximum amplitude. At this instant it has a very high rate of change, so the output of the differentiator is very high (see waveform in Fig. 12-3 at T_1). However, then the amplitude of the input signal reaches maximum and remains constant until T_2, when it drops back to its previous value. Thus, the differentiator will produce a sharp positive-going spike at T_1 and a sharp negative-going spike at T_2. In an ideal circuit there is no transition between these states, but in real circuits there is an exponential transition that is proportional to the RC time constant of the circuit and the rise time of the waveform. Differentiator output spikes are frequently used in circuits such as timers and zero-crossing detectors.

If a sine wave is applied to the inputs of either integrators or differentiators, then the result is a sine-wave output that is shifted in phase 90°. The principal difference between the two forms of circuit is in the direction of the phase shift. Such circuits are frequently used to provide quadrature or "sine-cosine" outputs from a sine-wave oscillator.

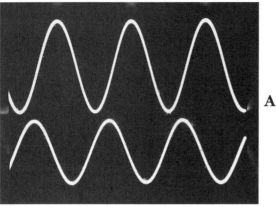

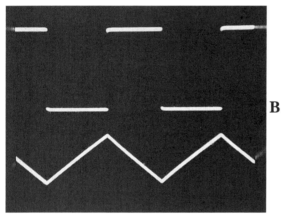

12-2 A. Effect of integrator on sine wave. B. On square wave.

RC integrator circuits

The simplest form of integrator and differentiator are simple resistor and capacitor circuits. The integrator is shown in Fig. 12-4, while the differentiator is in Fig. 12-5. The integrator consists of a resistor element in series with the signal line, and a capacitor across the signal line. The differentiator is just the opposite; the capacitor is in series with the signal line while the resistor is in parallel with the line. These circuits are also known as low-pass and high-pass RC filters, respectively. The low-pass case (integrator) has a –6 dB/octave falling-characteristic frequency response, while the high-pass case (differentiator) has a +6 dB/octave rising-characteristic frequency response.

The operation of the integrator and differentiator is dependent upon the time constant of the RC network (i.e., $R \times C$). The integrator time constant is set long (i.e., >10×) compared with the period of the signal being integrated, while in the differentiator the RC time constant is short (i.e., < ¹⁄₁₀×) compared with the period of the signal. Several integrators can be connected in cascade in order to increase the time-averaging effect, or increase the slope of the frequency response falloff.

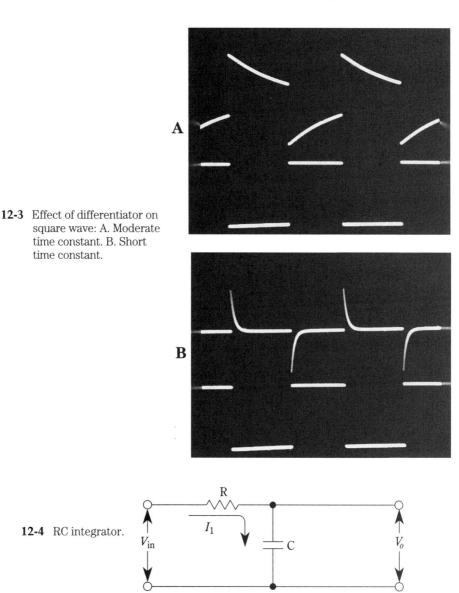

12-3 Effect of differentiator on square wave: A. Moderate time constant. B. Short time constant.

12-4 RC integrator.

Active differentiator and integrator circuits

The operational amplifier makes it relatively easy to build high-quality active integrator and differentiator circuits. Previously, one had to construct a stable, drift-free, high-gain transistor amplifier for this purpose. Figure 12-6 shows the basic circuit of the operational-amplifier differentiator. Again, the RC elements are used, but in a slightly different manner. The capacitor is in series with the op amp's inverting input, while the resistor is the op-amp feedback resistor.

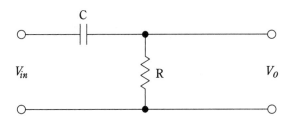

12-5 RC differentiator.

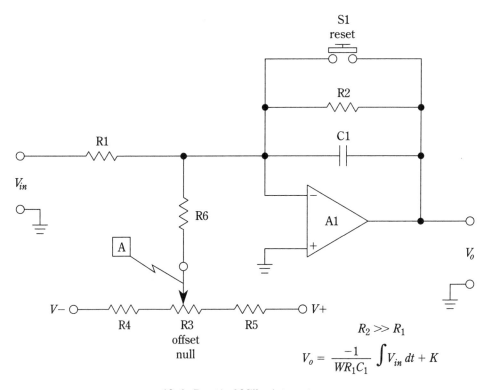

$$R_2 \gg R_1$$

$$V_o = \frac{-1}{WR_1C_1} \int V_{in}\, dt + K$$

12-6 Practical Miller integrator.

Analysis of the circuit to derive the transfer function follows a procedure similar to that followed for inverting and noninverting followers earlier in this book. From Kirchhoff's Current Law (KCL):

$$I_2 = I_1 \tag{12-1}$$

From basic passive circuit theory, including Ohm's law:

$$I_1 = \frac{C\Delta V_{in}}{\Delta t} \tag{12-2}$$

$$I_2 = \frac{V_o}{R} \tag{12-3}$$

Substituting (12-2) and (12-3) into (12-1):

$$\frac{V_o}{R} = \frac{-C\Delta V_{in}}{\Delta t} \tag{12-4}$$

or, with the terms rearranged:

$$V_o = -RC\frac{\Delta V_{in}}{\Delta t} \tag{12-5}$$

Where: V_o and V_{in} are in the same units (volts, millivolts, etc.)
R is in ohms
C is in farads
t is in seconds.

Equation 12-5 is a mathematical way of saying that output voltage V_o is equal to the product of the RC time constant times the derivative of input voltage V_{in} with respect to time (dV_{in}/dt). Since the circuit is essentially a special case of the familiar inverting follower circuit, the output is inverted, hence the negative sign.

Figure 12-7 shows the classical operational-amplifier version of the Miller integrator circuit. Again, an operational amplifier is the active element, while a resistor is in series with the inverting input and a capacitor is in the feedback loop. Notice that the placement of the capacitor and resistor elements are exactly opposite in both the RC and operational-amplifier versions of the integrator and differentiator circuits. In other words, the RC elements reverse roles between Figs. 12-7 and 12-8. That fact will tell the astute student quite a bit regarding the nature of integration and differentiation.

The output of the integrator is dependent upon the input signal amplitude and the RC time constant. The transfer function for the Miller integrator is derived in a manner similar to that of the differentiator:

From KCL:

$$I_2 = I_1 \tag{12-6}$$

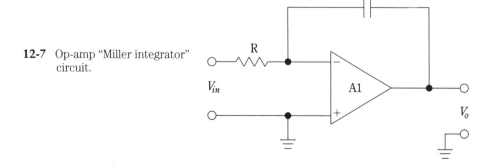

12-7 Op-amp "Miller integrator" circuit.

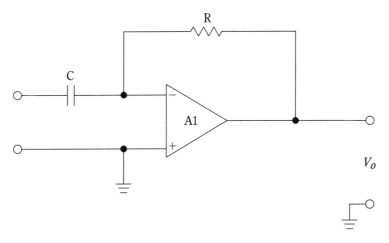

12-8 Op-amp differentiator.

From Ohm's law:

$$I_1 = \frac{V_{in}}{R}$$

(12-7)

and,

$$I_2 = C\,\frac{dV_o}{dt}$$

(12-8)

Substituting Eqs. 12-7 and 12-8 into Eq.12-6:

$$\frac{CdV_o}{dt} = \frac{-V_{in}}{R}$$

(12-9)

Integrating both sides:

$$\int \frac{CdV_o}{dt} = \int \frac{-V_{in}}{R}\,dt$$

(12-10)

$$VV_o = \int \frac{-V_{in}}{R}\,dt$$

(12-11)

Collecting and rearranging terms:

$$CV_o = \frac{-1}{R}\int V_{in}\,dt$$

(12-12)

$$V_o = \frac{-1}{RC}\int V_{in}\,dt$$

(12-13)

Then accounting for initial conditions:

$$V_o = \frac{-1}{RC} \int V_{in} \, dt + K \qquad\qquad \textbf{(12-14)}$$

Where: V_o and V_{in} are in the same units (volts, millivolts, etc.),
R is in ohms
C is in farads
t is in seconds.

This expression is a way of saying that the output voltage is equal to the time average of the input signal plus some constant K, which is the voltage that may have been stored in the capacitor from some previous operation (often zero in electronic applications).

Practical circuits

The circuits shown in Figs. 12-6 and 12-9 are classic textbook circuits. Unfortunately, they don't work very well in some practical cases. The problem is that these circuits are too simplistic because they depend upon the properties of ideal operational amplifiers. Unfortunately, real op amps fall far short of the ideal in several important ways that affect these circuits.

In real circuits, differentiators may "ring" or oscillate, and integrators may saturate from their tendency to integrate bias currents and other inherent dc offsets very shortly after turn-on.

There is another problem with this kind of circuit, and it magnifies the problem of saturation. Namely, the integrator circuit of Fig. 12-9 has a very high gain with certain values of R and C. The voltage gain of this circuit is given by the term $-1/RC$ which, depending on the values selected for R and C, can be quite high.

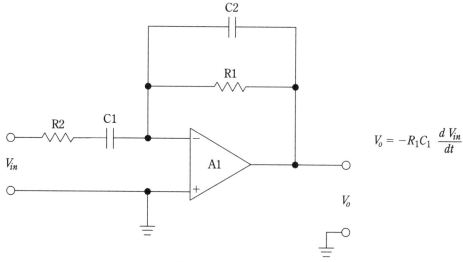

$$V_o = -R_1 C_1 \frac{d V_{in}}{dt}$$

12-9 Practical Circuit.

In other words, with a gain of −10,000, +1 V applied to the input will want to produce a −10,000-V output. Unfortunately, the operational-amplifier output is limited to the range of approximately −10 V to −20 V, depending upon the device and the applied V− dc power-supply voltage. For this case, the operational amplifier will saturate very rapidly! In order to keep the output voltage from saturating, it is necessary to prevent the input signal from rising too high. If the maximum output voltage allowable is 10 V, then the maximum input signal is 10 V/10,000 or 1 mV! Obviously, it is necessary to keep the RC time constant within certain bounds.

How to solve the problem Fortunately, some design tactics allow keeping the integration aspects of the circuit, while removing the problems. A practical integrator is shown in Fig. 12-6. The heart of this circuit is an RCA BiMOS operational amplifier, type CA-3140, or an equivalent BiFET device. The reason why this works so well is that it has a low input bias current (being MOSFET input).

Capacitor C1 and resistor R1 in Fig. 12-6 form the integration elements, and are used in the transfer equation to calculate performance. Resistor R2 is used to discharge C1 to prevent dc offsets on the input signal and the op amp itself from saturating the circuit. The reset switch is used to set the capacitor voltage back to zero (to prevent a "K" factor offset) before the circuit is used. In some measurement applications, the circuit is initialized by closing S1 momentarily. In actual circuits, S1 may be a mechanical switch, an electromechanical relay, a solid-state relay, or a CMOS electronic switch.

If there is still a minor drift problem in the circuit, then potentiometer R5 can be added to the circuit to cancel it. This component adds a slight counter current to the inverting input through resistor R6. To adjust this circuit, set $R5$ initially to midrange. The potentiometer is adjusted by shorting the V_{in} input to ground (setting $V_{in} = 0$), and then measuring the output voltage. Press S1 to discharge C1, and note the output voltage (it should go to zero). If V_o does not go to zero, then turn R5 in the direction that counters the change of V_o. This change can be observed after each time reset switch S1 is pressed. Keep pressing S1 and then making small changes in R5, until the setting is found at which the output voltage stays very nearly zero and constant, after S1 is pressed (there may be some very long-term drift).

Figure 12-9 shows the practical version of the differentiator circuit. The differentiation elements are R1 and C1, and the previous equation for the output voltage is used. Capacitor C2 has a small value (1 pF to 100 pF), and alters the frequency response of the circuit in order to prevent oscillation or ringing on fast rise-time input signals. Similarly, a "snubber" resistor (R2) in the input also limits this problem. The operational amplifier can be almost any type with a fast enough slew rate, and the CA-3140 is often recommended. The values of R2 and C2 are often determined by rule of thumb, but their justification is taken from the Bode plot of the circuit.

13

Triangle and sawtooth waveform generators

THE TRIANGLE AND SAWTOOTH WAVEFORMS (FIG. 13-1) ARE EXAMPLES OF *periodic ramp functions*. The *sawtooth* (Fig. 13-1A) is a single-ramp waveform. The voltage begins to rise linearly at time t_1 until it reaches a maximum. At time t_2, the waveform abruptly drops back to zero, where it again starts to ramp up linearly. The sawtooth is usually periodic, although single-sweep and triggered-sweep variants are sometimes seen. The period is designated T (see Fig. 13-1A); the frequency (f) is $1/T$.

The triangle waveform (Fig. 13-1B) is a double ramp. The waveform begins to ramp up linearly at time t_1. It reverses direction at time t_2 and then ramps downward linearly until time t_3. At time t_3, the waveform again reverses direction and begins ramping upwards. The period of the triangle waveform (T) is $T_1 - T_3$, and again frequency (f) is $1/T$, or $1/(T_1 - T_3)$.

Ramp generators are derived from capacitor charging circuits. The familiar RC charging curve was discussed in chapter 9, and is reproduced in simplified form in Fig. 13-1C. The RC charging waveform has an exponential shape, so is not well-suited to generating a linear ramp function.

There are two approaches to forcing the capacitor charging waveform to be more linear. The first is to limit the charging time to the short quasilinear segment shown in Fig. 13-1C. The ramp thus obtained is not very linear, is limited in amplitude to a small fraction of V_1, and has a relatively steep slope that may or may not be useful for any given application. A superior method is to charge the capacitor through a constant-current source (CCS). Using the CCS to charge the capacitor results in the linear ramp shown in Fig. 13-1C.

Triangle and sawtooth waveform oscillators create the constant-current form of ramp generator by using a Miller integrator circuit to charge the capacitor (Fig. 13-2A). When a Miller integrator is driven by a stable reference voltage source, the

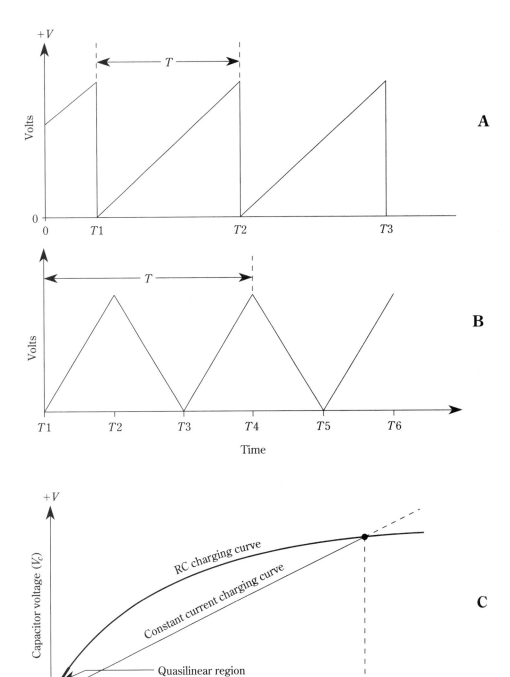

13-1   A. Sawtooth waveform. B. Triangle waveform. C. Generating ramp from capacitor charging waveforms.

output is a linearly rising ramp. The ramp voltage (V_o) is:

$$V_o = \frac{V_{ref}}{T} \qquad \textbf{(13-1)}$$

or, because $T = RC$:

$$V_o = \frac{V_{ref}}{RC} \qquad \textbf{(13-2)}$$

If V_{ref} = +10 Vdc, and the RC time-constant is $T = RC = 0.001$ s, the ramp slope is:

$$V_o = \frac{10 \text{ V}}{0.001 \text{ s}}$$

$$= 0.10 \text{ V}/s$$

Triangle generator circuits

Figure 13-2A shows a simplified circuit model of a triangle-waveform generator, while Fig. 13-2B shows the expected waveforms. This circuit consists of a Miller integrator as the ramp generator, and an SPDT switch (S1) that can select either positive ($+V_{ref}$) or negative ($-V_{ref}$) reference voltage sources. For purposes of this discussion, switch S1 is an electronic switch that is toggled back and forth between positions A and B by a square wave applied to the control terminal (CT). Assume an initial condition at time t_2 (at which point $V_o = -V_1$) and the input of the integrator is connected to $-V_{ref}$. At time t_2, the square-wave switch driver changes to the opposite state, so S1 toggles to connect $+V_{ref}$ to the integrator input. The ramp output will rise linearly at a rate of $+V_{ref}/RC$ until the switch again toggles at time t_3. At this point, the ramp is under the influence of $-V_{ref}$, so drops linearly from $+V_1$ to $-V_1$. The switch continuously toggles back and forth between $-V_{ref}$ and $+V_{ref}$, so the output (V_o) continously ramps back and forth between $-V_1$ and $+V_1$.

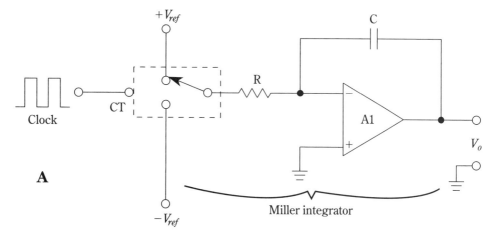

13-2A Notional triangle waveform generator.

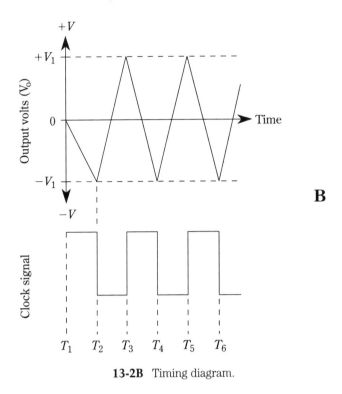

13-2B Timing diagram.

The circuit of Fig. 13-2A is not really practical, but serves as an analogy for the actual circuit. Figure 13-3A shows the circuit for a practical triangle-waveform generator in which a Miller integrator forms the ramp generator and a voltage comparator serves as the switch (see also the waveforms in Fig. 13-3B). The comparator uses the positive-feedback configuration, so operates as a noninverting Schmitt trigger. Such a circuit snaps high ($V_B = +V_{sat}$) when the input signal crosses a certain threshold voltage in the positive-going direction. It will snap low again ($V_B = -V_{sat}$) when the input signal crosses a second threshold in a negative-going direction. The two thresholds are not always the same potential.

Because zener diodes D1 and D2 are in the circuit, the maximum allowable value of $+V_B$ is $[V_{ZD1} + 0.7]$ V, while the limit for $-V_B$ is $-[V_{ZD2} + 0.7]$ V. If $V_{ZD1} = V_{ZD2}$, then $|+V_B| = |-V_B|$. These potentials represent $\pm V_{ref}$ discussed in the analogy presented previously, so are the potentials that become the ramp-generator input signal.

Consider an initial state in which V_B is at the negative limit $-V_B$. The output V_o will begin to ramp upwards from a minimum voltage of:

$$V_1 = \left(\frac{V_a(R_2 + R_4)}{R_4} \right) - \left(\frac{V_B R_2}{R_4} \right) \tag{13-3}$$

The output will continue to ramp upwards towards a maximum value of:

$$V_3 = \left(\frac{V_A(R_2 + R_4)}{R_4} \right) + \left(\frac{V_B R_2}{R_4} \right) \tag{13-4}$$

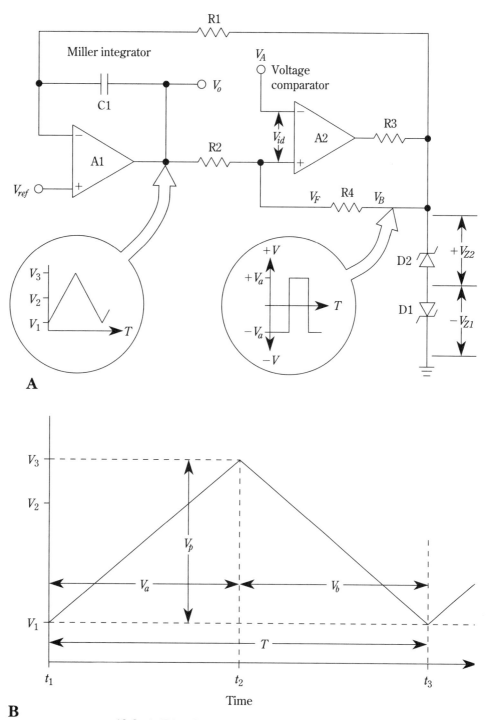

A

B

13-3 A. Triangle waveform generator circuit; B. Timing waveform.

Causing a peak swing voltage of:

$$V_p = V_3 - V_1 \tag{13-5}$$

$$= \left(\left(\frac{V_A(R_2 + R_4)}{R_4} \right) + \left(\frac{V_B R_2}{R_4} \right) \right) - \left(\left(\frac{V_A(R_2 + R_4)}{R_4} \right) - \left(\frac{V_B R_2}{R_4} \right) \right)$$

$$= \frac{V_B R_2}{R_4} + \frac{V_B R_2}{R_4}$$

$$V_p = \frac{2V_B R_2}{R_4} \tag{13-6}$$

Switching of the comparator occurs when the differential input voltage V_{id} is zero. The inverting input (−IN) voltage is V_A, which is a fixed reference potential. The noninverting input (+IN) is at a voltage (V_F) that is the superposition of two voltages, V_o and V_B:

$$V_F = \left(\frac{V_o R_4}{R_2 + R_4} \right) + \left(\frac{\pm V_B R_2}{R_2 + R_3} \right) \tag{13-7}$$

If $+V_B = -V_B$, then the positive and negative thresholds are equal.
The duration of each ramp (t_a and t_b) can be found from:

$$t_{a,b} = \frac{V_p}{\left(\dfrac{V_B}{R_1 C_1} \right)} \tag{13-8}$$

The value of V_B is selected from $-V_B$ or $+V_B$ as needed. In Eq. 13-6 it was found that $V_p = 2V_B R_2/R_4$, so:

$$t_{a,b} = \frac{\left(\dfrac{2V_B R_2}{R_4} \right)}{\left(\dfrac{V_B}{R_1 C_1} \right)}$$

$$= \left(\frac{R_1 C_1}{V_B} \right) \left(\frac{2V_B R_2}{R_4} \right)$$

$$t_{a,b} = R_1 C_1 \left(\frac{2V_B R_2}{R_4} \right) \tag{13-9}$$

or, in the less general (but more common) case of $t_a = t_b$:

$$T = 2R_1 C_1 \left(\frac{2R_2}{R_4} \right) \tag{13-10}$$

The frequency (f) of the triangle wave is the reciprocal of the period ($1/T$), so:

$$f = \frac{1}{T} \qquad\qquad (13\text{-}11)$$

$$= \frac{1}{\left(\dfrac{4R_1 C_1 R_2}{R_4} \right)}$$

$$= \frac{R_4}{4R_1 C_1 R_2}$$

Experiment 13-1

Experiment 13-1 examines the use of a Miller integrator circuit to generate a triangle waveform. The Miller integrator is an operational-amplifier circuit in which a capacitor is used as the negative-feedback element. When a dc voltage is applied to the input of the integrator circuit, the output voltage, V_o, rises in a ramp-function manner.

1. Connect the circuit of Fig. 13-4. Use either a CA-3140, which is a BiMOS op amp, or an equivalent BiFET op amp (we are looking for a device with a minimal input bias current). Use a pair of 1.5-V "AA" cells for the dc signal source. Connect either a dc-coupled oscilloscope or a voltmeter to the output. If the voltmeter is an analog type, then it should be set to display 0 V at the center of the scale.

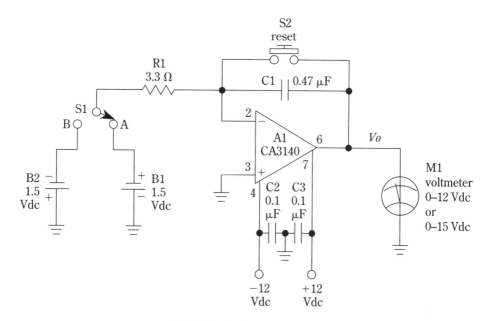

13-4 Circuit for experiment 13-1.

2. The gain of the Miller integrator of Fig. 13-4 is $-1/R_1C_1$, where R_1 is in ohms and C_1 is in farads. Calculate the gain for the circuit shown. You should get an answer of 1.55.
3. Set switch S1 to position "A," and then turn on the power to the circuit.
4. Press switch S2 to discharge the capacitor. Immediately after releasing S2, watch the output indicator (scope or voltmeter). It should begin rising.
5. When V_o gets to about 10 V, change switch S1 to position "B" and observe what happens to the output voltage.

Sawtooth generators

The *sawtooth wave* (Fig. 13-1A) is a single-slope ramp function. The wave ramps linearly upward (or downward), and then abruptly snaps back to the initial base-line condition. Figure 13-5A shows a simple model of a sawtooth-generator circuit. A constant-current source charges a capacitor in a manner that generates the linear ramp function (Fig. 13-5B). When the ramp voltage (V_c) reaches the maximum point

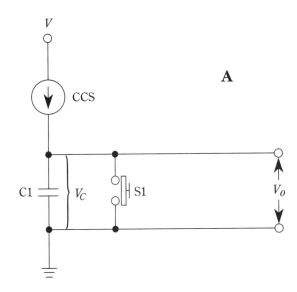

13-5   A. Notional sawtooth generator. B. Timing waveform.

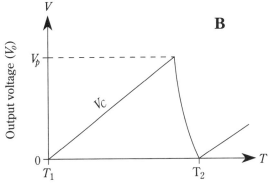

(V_p), switch S1 is closed, forcing V_c back to zero by discharging the capacitor. If switch S1 remains closed, then the sawtooth is terminated. If S1 reopens, however, then a second sawtooth is created as the capacitor recharges.

Figure 13-6A shows the circuit for a periodic sawtooth oscillator. It is similar to Fig. 13-5A, except that a junction field-effect transistor (JFET), Q1, is used as the discharge switch. When Q1 is turned off, the output voltage ramps upward (see Fig. 13-6B). When the gate is pulsed hard on, the drain-source channel resistance drops from a very high value to a very low value, forcing C1 to discharge rapidly. In the absence of a gate pulse, however, the channel resistance remains very high. At time t_1, the gate is turned

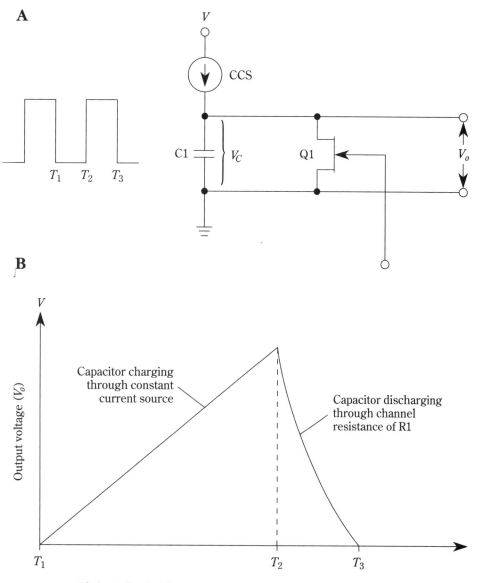

13-6 A. Clock-driven sawtooth generator. B. Timing diagram.

off, so V_c begins ramping upward. At t_2, the JFET gate is pulsed, so C1 rapidly discharges back to zero. When the pulse $(t_2 - t_3)$ ends, however, Q1 turns off again and the ramp starts over. The same circuit can be used for single-sweep operation by replacing the pulse train applied to the gate of Q1 with the output of a monostable multivibrator.

The circuit of Fig. 13-7A shows a sawtooth generator that uses a Miller integrator (A1) as a ramp generator, and replaces the discharge switch with an electronic switch that is driven by a voltage comparator and one-shot circuit. The timing diagram for this circuit is shown in Fig. 13-7B. Under the initial conditions, at time t_1, the output voltage (V_o) ramps upward at a rate of $[-(-V_{ref})/R_1C_1]$. The voltage comparator (A2) is biased with the noninverting input (+IN) set to V_1, and the inverting input at V_o. The comparator differential input voltage $V_{id} = (V_1 - V_o)$. As long as $V_1 > V_o$, the comparator sees a negative input, so produces a HIGH output of $+V_{sat}$. At the point where $V_1 = V_o$, the differential input voltage is zero, so the output of A2 (voltage V_2) drops low (i.e., $-V_{sat}$). The negative-going edge of V_2 at time t_2 triggers the one-shot circuit. The output of the one-shot briefly closes electronic switch S1, causing the capacitor to discharge. The one-shot pulse ends at time t_3, so S1 reopens and allows V_o again to ramp upward.

Project 13-1   Digitally generated sawtooth and triangle waveforms

Sawtooth signal generators are used for a variety of purposes in electronics: electronic music synthesizers, sweeping RF, audio signal generators, voltage-controlled oscillators (VCO), certain bench tests, calibrating oscilloscopes, and other applica-

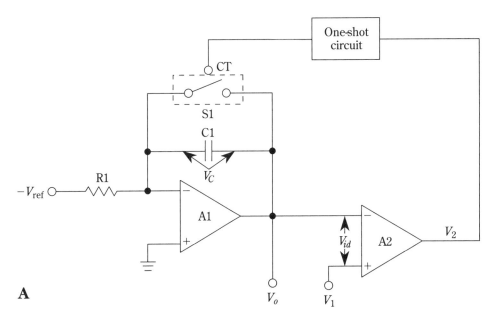

A

13-7A Sawtooth generator circuit.

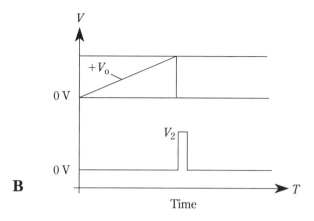

13-7B Timing diagram.

tions. In addition, many circuit applications exist for embedded sawtooth generators. Examples include many situations in which the sawtooth precisely calibrates an oscilloscope timebase. If the sawtooth that sweeps the oscilloscope horizontally is controlled from a stable crystal oscillator, then a very precise sweep rate is possible. The "standard" solid-state sawtooth generator circuit consists of an operational-amplifier Miller integrator circuit excited by a square wave. However, there are problems with such circuits.

Figure 13-8 shows part of the problem with the standard op-amp sawtooth-generator circuit. The waveform has two principal defects. First, the ramp-up edge (T_1-T_2) of the "sawtooth" is not linear. Because the original design uses a capacitor charge/discharge circuit in the Miller integrator, the ramp naturally has a shape like the normal capacitor-charge waveform. What is required of a proper sawtooth is a linear ramp—i.e., one that rises as a straight line.

The second defect with the waveform in Fig. 13-8 is the fall time ($T_2 - T_3$): it's too long. Although proper design will make the Miller integrator sawtooth generator closer to the ideal, the use of a few low-cost digital components produces a more perfect sawtooth generator without headaches.

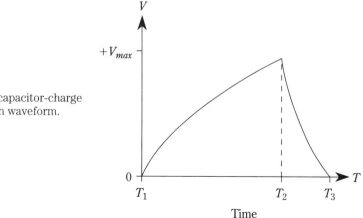

13-8 Typical capacitor-charge sawtooth waveform.

The circuit for a digitally synthesized sawtooth generator is shown in Fig. 13-9A. The heart of this circuit is IC1, a DAC0806 eight-bit digital-to-analog converter (DAC). This DAC is based on the MC-1408 family of DACs, and was selected because it is well-behaved in simple, easy-to-use circuitry, and is easily available through mail-order sources such as Jameco Electronics or in blister packs through local distributers.

A DAC produces an output current that is proportional to the reference voltage or current and to the binary word applied to the digital inputs. The controlling function for the DAC selected for this article is of the form:

$$I_O = I_{ref}\left(\frac{A}{256}\right) \tag{13-12}$$

Where: I_O is the output current from pin 4
  I_{ref} is the reference current applied to pin 14
  A is the decimal value of the binary word applied to the eight binary inputs
    (pins 5–12)

The reference current is found from Ohm's law, and is the quotient of the reference voltage and the series resistor at pin 14 ($R4$). In data-acquisition systems (the main use for the DAC), the reference voltage is a precision, regulated potential. But in this case we do not need the precision, so use the V+ power supply as the reference voltage. Therefore, the reference current is (+12 Vdc)/R_4. With the value of R_4 shown (6800 Ω), I_{ref} is 0.00018 A, or 1.8 mA. Values from 500 μA to 2 mA are permissible with this device. If you elect to change the reference current, be sure to keep $R_4 = R_5$.

The reference current sets the maximum value of output current, I_O. When a full-scale binary word (11111111) is applied to the binary inputs, the output current I_O is:

$$I_O = (1.8 \text{ mA})\left(\frac{255}{256}\right) \tag{13-13}$$
$$= (1.8 \text{ mA})(0.996)$$
$$= 1.78 \text{ mA}$$

The DAC0806 is a current-output DAC, so we must use an op-amp current-to-voltage converter to make a sawtooth voltage function. Such a circuit is an ordinary inverting follower without an input resistor. The output voltage (V_O) will rise to a value of ($I_O R_3$).

The actual output waveform is "staircased" (Fig. 13-9B) in binary steps equal to the 1-LSB current of IC1 (or the 1-LSB voltage of V_O). The 1-LSB voltage is the smallest step change in output potential caused by flipping the least-significant bit (B1) either from 0 to 1 or from 1 to 0. The reason why you don't see the steps is that the frequency response of the 741 operational amplifier used for the current-to-voltage converter acts as a low-pass filter to smooth the waveform. If a higher frequency op amp is used, then a capacitor shunting R_3 will serve to low-pass filter the waveform. A –3 dB frequency (f) of 1 or 2 kHz will suffice to smooth the waveform. The value of the capacitor is calculated from:

$$C_{\mu F} = \frac{1,000,000}{2\pi R_3 f} \tag{13-14}$$

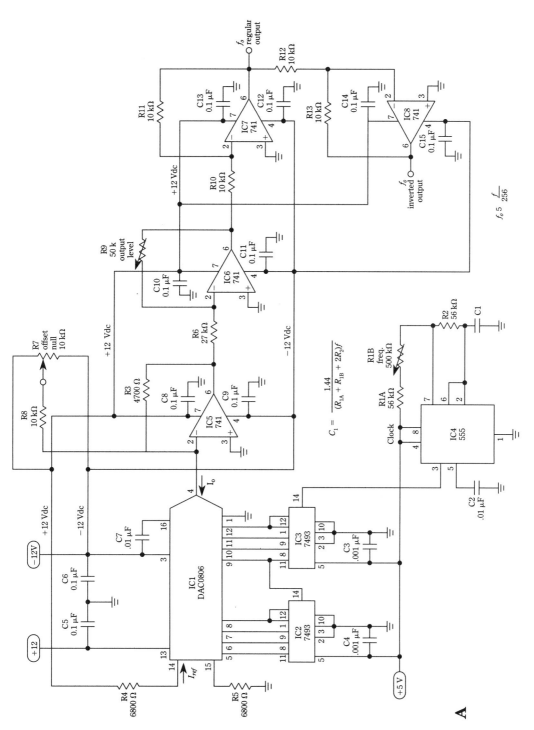

13-9A Digital sawtooth generator circuit.

$$C_1 = \frac{1.44}{(R_{1A} + R_{1B} + 2R_2)f}$$

$$f_o, 5\ \frac{f}{256}$$

A

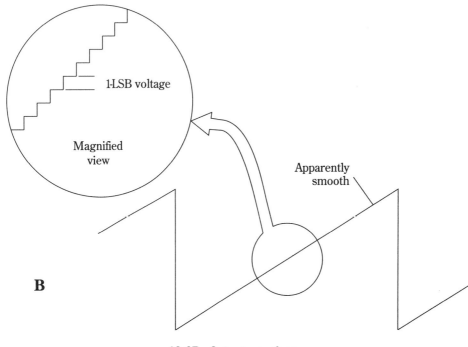

1-LSB voltage

Magnified view

Apparently smooth

B

13-9B Output waveform.

Where: $C_{\mu F}$ is the capacitance in microfarads
f is the –3 dB cut-off frequency in hertz (Hz)
R_3 is expressed in ohms

This circuit is synchronized by a clock oscillator consisting of a single 555 IC timer. Although not a TTL device, the 555 is TTL-compatible when the V+ potential applied to pins 4 and 8 is limited to +5 Vdc. The 555 is connected in the astable multivibrator configuration, so it outputs a chain of pulses with a +4-V amplitude. The operating frequency is set by three resistors (R1, R2, and R6) and a capacitor (C7). The actual clock frequency is:

$$f = \frac{1.44}{((R_3 + R_{12}) + 2R_4)C} \tag{13-15}$$

Where: f is the frequency in hertz (Hz)
C is the value of C7 in farads
$R_1, R_2,$ and R_6 are in ohms

Select a clock frequency 256 times the desired sawtooth fundamental frequency.

Generating other waveforms

As electronic music buffs will testify, you can get almost any waveform you need by applying the right binary words to the digital inputs of the DAC0806. Because I wanted a sawtooth waveform, the DAC digital inputs were connected to the outputs of an 8-bit binary counter built from a pair of 7493 TTL base-16 counter chips. Each chip is a 4-bit counter, so they are cascaded to produce the 8-bit binary word needed to drive the DAC. If you want a detailed description of this chip, then I recommend Don Lancaster's book *TTL Cookbook*. The function of this counter is to increment in steps from 00000000 to 11111111 under control of a clock signal applied to the input (pin 14) of IC2. You could use any 8-bit counter that outputs a TTL-compatible signal in place of the 7493 devices that I selected. The 7493 was selected for the best of all engineering reasons—I had a pair of them in my junkbox.

If you want a triangle waveform, then it is possible to replace the 7493 devices with a base-16 up/down counter chip. Arrange the digital control logic to reverse the direction of the count when the maximum state (11111111) is sensed.

There are two ways to generate waveforms other than a sawtooth or triangle, and both of them involve using a computer memory. The binary bit pattern representing the waveform is stored in memory, and then output in the right sequence. One method uses a read-only memory (ROM), which you preprogram with the bit pattern representing the waveform. A binary-counter circuit connected as an address generator selects the bit-pattern sequence.

The second method for generating waveforms is to store the bit pattern in a computer, and then output it under program control via an 8-bit parallel-output port. This method is usable both for generating special waveforms and for linearizing the tuning characteristic of circuits such as VCOs, swept oscillators, etc. The digital solution to the linearization problem involves storing a look-up table in either a ROM or computer memory. I learned this system in a laboratory where it was once used to linearize low-level pressure-transducer measurements. Interfacing a computer output (assuming that you have an 8-bit parallel port available) is simple. Connect the output of the computer directly to the input of the DAC.

The digitally synthesized sawtooth generator is easy to build, and is well-behaved—so it shouldn't haunt you too much in the building. In addition, it is easy to modify to generate any waveform. You can expect to see more direct digitally synthesized circuits in test equipment and communications equipment in the future.

14

Sine wave audio oscillators

AS EXPLAINED IN CHAPTER 7, FEEDBACK OSCILLATOR (FIG. 14-1) CONSISTS OF AN amplifier with an open-loop gain of A_{vol} and a feedback network with a gain or transfer function β. It is called a "feedback oscillator" because the output signal of the amplifier is fed back to the amplifier's own input by way of the feedback network. Figure 14-1 is a block diagram model of the feedback oscillator.

Sine wave oscillators

Sine wave oscillators produce an output signal that is sinusoidal. Such a signal is ideally very pure, and if indeed it is perfect, then its Fourier spectrum will contain only the fundamental frequency and no harmonics. It is the harmonics in a nonsinusoidal waveform that give it a characteristic shape. The active element in the circuits described in this circuit is the operational amplifier. However, any linear amplifier will also work in place of the operational amplifier. The one circuit that shows the principles most clearly is the RC phase-shift oscillator, so this discussion starts with that circuit.

Stability in oscillator circuits can refer to several different phenomena. First is *frequency stability* (see chapter 19), which refers to the ability of the oscillator to remain on the design frequency over time. Several different factors affect frequency stability, but the most important are temperature and power-supply voltage variations. Another form of stability is *amplitude stability*. Because sine-wave oscillators do not operate in the saturated mode, it is possible for minor variations in overall circuit gain to affect the amplitude of the output signal. Again, the factors most often cited for this problem include temperature and dc power-supply variations. The latter is overcome by using regulated dc power supplies for the oscillator. The former is overcome by either temperature-compensated design or maintaining a constant operating temperature. Some variable sine-wave oscillators will exhibit amplitude

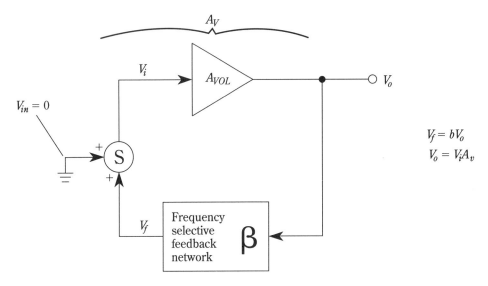

$$V_f = bV_o$$
$$V_o = V_iA_v$$

14-1 Feedback oscillator circuit block diagram.

variation of the output signal when the operating frequency is changed. In these circuits either a self-compensation element is used, or an automatic-level-control amplifier stage is used.

Still another form of stability regards the purity of the output signal. If the circuit exhibits spurious oscillations, then these will be superimposed on the output signal. As with any circuit containing an op amp or any other high-gain linear amplifier, it is necessary to properly decouple the dc power supply lines. It may also be necessary to frequency-compensate the circuit.

RC phase-shift oscillator circuits

The RC phase-shift oscillator is based on a three-stage cascade resistor-capacitor network such as shown in Fig. 14-2A. An RC network exhibits a phase-shift ϕ (Fig. 14-2B) that is a function of resistance (R) and capacitive reactance (X_c). Because Xc is inversely proportional to frequency ($1/2\pi fC$), the phase angle shift across the network is therefore a function of frequency.

The goal in designing the RC phase-shift oscillator is to create a phase shift of 180° between the input and output of the network at the desired frequency of oscillation. It is conventional practice to make the three stages of the network identical, so that each provides a 60° phase shift. Although it is common practice, it is also not strictly necessary, provided that the total phase shift is 180°. One reason for using identical stages, however, is that it is possible for the nonidentical designs to have more than one frequency for which the total phase shift is 180°. This phenomenon can lead to undesirable multimodal oscillation.

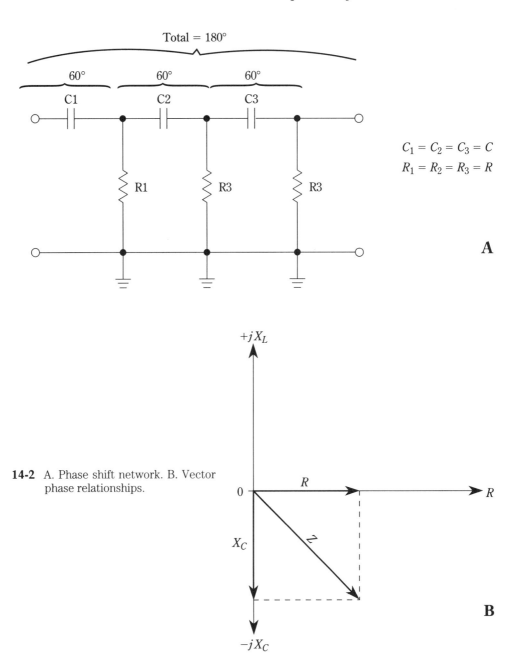

14-2 A. Phase shift network. B. Vector phase relationships.

Figure 14-3 shows the circuit for an operational-amplifier RC phase-shift oscilla-tor. The cascade phase-shift network $R_1R_2R_3/C_1C_2C_3$ provides 180° of phase shift at a specific frequency, while the amplifier provides another 180° (because it is an in-verting follower). The total phase shift is therefore 360° at the frequency for which the RC network provides a 180° phase shift. The frequency of oscillation (f) for this circuit is given by:

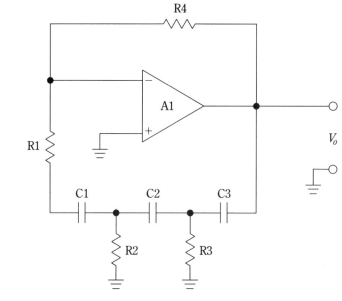

$R_1 = R_2 = R_3 = R$
$C_1 = C_2 = C_3 = C$
$R4 \geq 2qR$
$F_o = \dfrac{1}{2\pi\sqrt{6}\,RC}$

14-3 Op-amp RC phase-shift oscillator.

$$f = \frac{1}{2\pi\sqrt{6}\,RC} \qquad\qquad\text{(14-1)}$$

Where: f is in hertz (Hz)
$\quad$ R is in ohms
$\quad$ C is in farads

It is common practice to combine the constants in Eq.[14-1] to arrive at a simplified expression:

$$f = \frac{1}{15.39RC} \qquad\qquad\text{(14-2)}$$

Because the required frequency of oscillation is usually determined from the application, it is necessary to select an RC time constant to force the oscillator to operate as needed. Also, because capacitors come in fewer standard values, it is common practice to select an arbitrary trial value of capacitance, and then select the resistance that will cause the oscillator to produce the correct frequency. Also, to make the calculations simpler, it is prudent to express the equation in such a way that permits specifying the capacitance (C) in microfarads. As a result Eq. 14-2 is sometimes rewritten as:

$$R = \frac{1,000,000}{15.39 C_{\mu F} f} \qquad\qquad\text{(14-3)}$$

The attenuation through the feedback network must be compensated by the amplifier if loop gain is to be unity or greater. At the frequency of oscillation, the attenuation is 1/29. The loop gain must be unity, so the gain of amplifier A1 must be at least 29 in order to satisfy $A\beta = 1$. For the inverting follower (as shown), $R_1 = R$, and $A_v = R_4/R_1$. Therefore, it can be concluded that $R_4 \geq 29R$ in order to meet Barkhausen's criterion for loop gain.

Design example 14-1

Design a 1000-Hz sine-wave oscillator based on the RC phase-shift circuit of Fig. 14-3. Select values for R, C, and the feedback resistor.

1. Select trial value for C: 0.01 μF is a good start.
2. Solve Eq. 14-2 for R as a function of f:

$$R = \frac{1}{15.39fC}$$

$$= \frac{1}{(15.39)\,(1000 \text{ Hz})\,C}$$

$$= \frac{1}{15{,}390\,C}$$

3. Substitute the first trial value for C:

$$R_{(@1000 \text{ Hz})} = \frac{1}{(15{,}390)\,(0.01 \times 10^{-6}\text{F})}$$

$$= \frac{1}{1.539 \times 10^{-4}}$$

$$= 6498 \ \Omega$$

4. The value of R is 6498 Ω, which can be obtained as a 1% precision resistor.
5. Select a minimum value for R_4:

$$R_4 \geq 29\,R$$
$$R_4 \geq (29)\,(6498)$$
$$\geq 188{,}434 \ \Omega$$

6. A practical value of R_4 is greater than the calculate value, such as 200 kΩ or 220 kΩ. It should not be too much higher, however, or stability problems may arise.

Experiment 14-1

Build the sine-wave oscillator discussed in the previous design example. Use a two-channel oscilloscope to examine the following:

1. The output voltage, V_o.
2. The signal applied to the inverting input of the op amp. This step may require disconnecting the –IN end of R_1.

Wien-bridge oscillator circuits

The Wien-bridge circuit is shown in Fig. 14-4. Like several other well-known bridge circuits, the Wien bridge consists of four impedance arms. Two of the arms (R_1, R_2) form a resistive voltage divider that produces a voltage V_1 of:

$$V_1 = \frac{V_{ac}R_4}{R_3 + R_4} \tag{14-4}$$

The remaining two arms (Z_1, Z_2) are complex RC networks that consist of one capacitor and one resistor each. Impedance Z_1 is a series RC network, while Z_2 is a parallel RC network. The voltage and phase shift produced by the Z_1/Z_2 voltage divider are functions of the RC values and the applied frequency. Note that $V_2 = V_{ac}/3$, and that V_1 and V_{ac} are in-phase with each other.

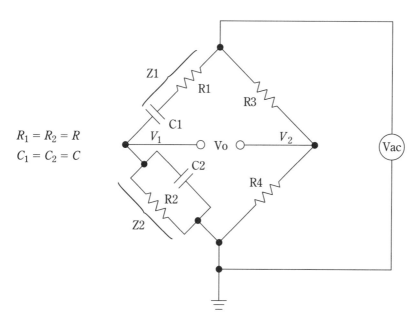

$R_1 = R_2 = R$
$C_1 = C_2 = C$

14-4 Wein-bridge circuit.

Figure 14-5 shows the circuit for a Wien-bridge oscillator. The resistive voltage divider supplies V_1 to the inverting input (–IN), while V_2 is applied to the noninverting input (+IN). In Fig. 14-5, the bridge signal source is the output of the amplifier (A1). The ac signal is applied to +IN, so the gain it sees is found from:

$$A_v = \frac{R_3}{R_4} + 1 \tag{14-5}$$

The ac feedback applied to +IN is:

$$\beta = \frac{Z_2}{Z_1 + Z_2} \tag{14-6}$$

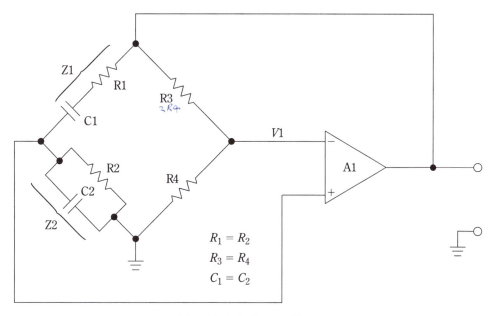

14-5 Wein-bridge oscillator.

At resonance, $B = \frac{1}{3}$, so (as shown in Fig. 14-4B):

$$V_2 = \frac{V_o}{3} \qquad (14\text{-}7)$$

Because $A_v = V_o/V_2$ by definition, satisfying Barkhausen's loop gain criterion $(-A_v\beta = 1)$ requires that $A_v = V_o/V_2 = 3$. Using this result:

$$A_v = \frac{R_3}{R_4} + 1 \qquad (14\text{-}8)$$

Or,

$$R_3 = 2R_4 \qquad (14\text{-}9)$$

If $R_1 = R_2 = R$ and $C_1 = C_2 = C$, the resonant frequency of the Wien bridge is:

$$f = \frac{1}{2\pi RC} \qquad (14\text{-}10)$$

For the standard Wien-bridge oscillator, in which $R_1 = R_2 = R$ and $C_1 = C_2 = C$, and $R_3 = 2R_4$, a sine-wave output will result on frequency f.

Design example 14-2

Design a 2000-Hz sine-wave oscillator using the Wien-bridge circuit of Fig. 14-5.

1. Select a trial value of capacitance: 0.01 μF.
2. Calculate the value of R by solving Eq. 14-10 for R when f is 2000 Hz.

$$R = \frac{1}{2\pi f C}$$

$$R = \frac{1}{(2)\,(3.14)\,(2000\ \text{Hz})\,(0.01 \times 10^{-6} F)}$$

$$= \frac{1}{1.256 \times 10^{-4}}$$

$$= 7961\ \Omega$$

3. The nearest standard value for R is 8200 Ω. If the exact frequency is not important, then select one of these units for the resistor. Otherwise, use a precision 1% resistor for R. Also, the precision resistor offers better frequency stability with temperature.

Experiment 14-2

Connect the Wien-bridge oscillator circuit using $C = 0.015\ \mu F$ and $R_1 = 15\ k\Omega$. Find the operating frequency, and observe the waveform on an oscilloscope. Try changing the values of C and R independently, while observing the effects on waveshape and frequency.

Amplitude stability

The oscillations in the Wien-bridge oscillator circuit want to build up without limit when the gain of the amplifier is high. Figure 14-6 shows the result of the gain being only slightly above that required for stable oscillation. Note that some clipping is beginning to appear on the sine-wave peaks. At even higher gains the clipping becomes more severe, and will eventually look like a square wave. Figure 14-7 shows several methods for stabilizing the waveform amplitude. Figure 14-7A shows the use of small-signal diodes such as the 1N914 and 1N4148 devices. At low signal amplitudes, the diodes are not sufficiently biased, so the gain of the circuit is:

$$A_v = \frac{R_1 + R_3}{R_2} \qquad \qquad \textbf{(14-11)}$$

As the output signal voltage increases, however, the diodes become forward biased. D1 is forward biased on negative peaks of the signal, while D2 is forward biased on positive peaks. Because D1 and D2 are shunted across R_3, the total resistance R_3' is less than R_3. By inspection of Eq. 14-11, you can determine that reducing R_3 to R_3' reduces the gain of the circuit. The circuit is thus self-limiting.

Another variant of the gain-stabilized Wien-bridge oscillator is shown in Fig. 14-7B. In this circuit, a pair of back-to-back zener diodes provide the gain-limitation function. With the resistor ratios shown, the overall gain is limited to slightly more than unity, so the circuit will oscillate. The output peak voltage of this circuit is set by the zener voltages of D1 and D2 (which should be equal for low-distortion operation).

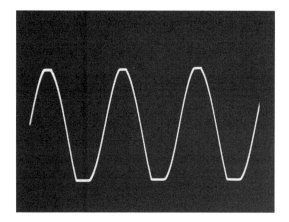

14-6 Clipping in the Wein-bridge oscillator.

One final version of the gain-stabilized oscillator is shown in Fig. 14-7C. In this circuit a small incandescent lamp is connected in series with resistor R2. When the amplitude of the output signal tries to increase above a certain level, the lamp will draw more current, causing the gain to reduce. The lamp-stabilized circuit is probably the most popular form where stable outputs are required. A thermistor is sometimes substituted for the lamp.

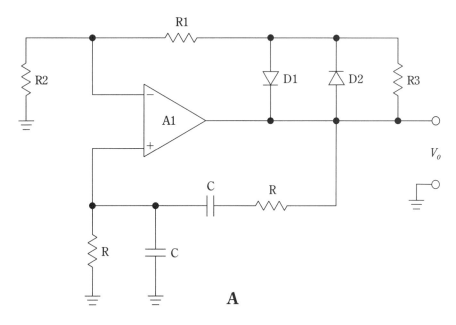

14-7A Improved circuit for limiting clipping.

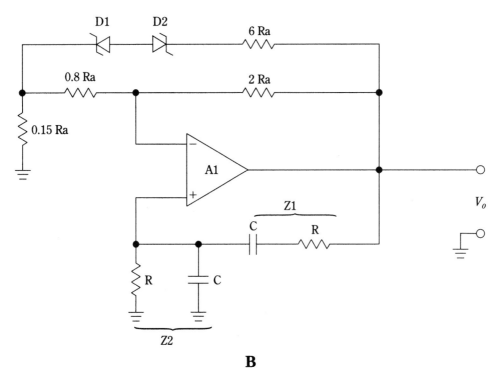

B

14-7B Zener diode version.

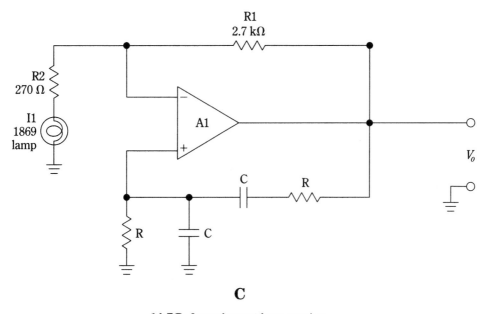

C

14-7C Incandescent lamp version.

Quadrature and biphasic oscillators

Signals that are "in quadrature" are of the same frequency but are phase shifted 90° with respect to each other. An example of quadrature signals is sine and cosine waves (Fig. 14-8A). Applications for the quadrature oscillator include demodulation of phase-sensitive detector signals in data-acquisition systems. The sine wave has an instantaneous voltage $v = V \sin(\omega_o t)$, while the cosine wave is defined by $v = V \cos(\omega_o t)$. Note that the distinction between sine and cosine waves is meaningless unless either both are present or some other timing method is used to establish when "zero degrees" is supposed to occur. Thus, when sine and cosine waves are called for, it is in the context of both being present, and a phase shift of 90° is present between them.

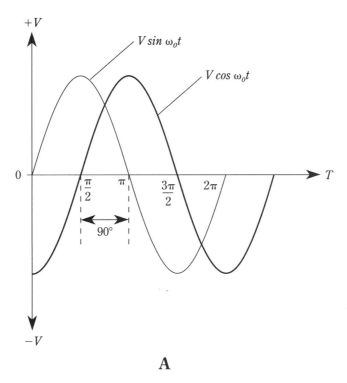

A

14-8A   Sine/cosine oscillator outputs.

The circuit for the quadrature oscillator is shown in Fig. 14-8B. It consists of two operational amplifiers, A1 and A2. Both amplifiers are connected as Miller integrators, although A1 is a noninverting type while A2 is an inverting integrator. The output of A1 (V_{o1}) is assumed to be the sine-wave output. In order to make this circuit operate, a total of 360° of phase shift is required between the output of A1, around the loop, and back to the input of A1. Of the required 360° of phase shift, 180° are provided by the inversion inherent in the design of A2 (it is in the inverting configuration). Another 90°

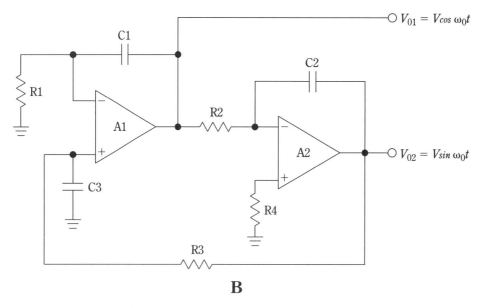

B

14-8B   Sine/cosine oscillator circuit (quadrature oscillator).

obtains from the fact that A2 is an integrator, which inherently causes a 90° phase shift. An additional 90° phase shift is provided by RC network R_3C_3. If $R_1 = R_2 = R_3 = R$, and $C_1 = C_2 = C_3 = C$, then the frequency of oscillation is given by:

$$f = \frac{1}{2\pi RC} \tag{14-12}$$

The cosine output (V_{o2}) is taken from the output of amplifier A2. The relative amplitudes are approximately equal, but the phase is shifted 90° between the two stages.

A *biphasic oscillator* is a sine wave oscillator that outputs two identical sine wave signals that are 180° out of phase with each other. The basic circuit is simple, and is shown in block-diagram form in Fig. 14-9. The biphasic oscillator consists of a sine-wave oscillator followed by a noninverting amplifier to produce the [$V \sin (\omega_o t)$ output, and an inverting amplifier that has a gain of −1 to produce the [$V \sin(\omega_o t + 2\pi)$] output. Two amplifiers are needed to keep the phase error at a minimum. Otherwise, the propagation time of a single amplifier would cause a slight delay in the $V \sin(\omega_o + 2\pi t)$ signal. Biphasic oscillators are sometimes used in transducer excitation applications in carrier amplifiers.

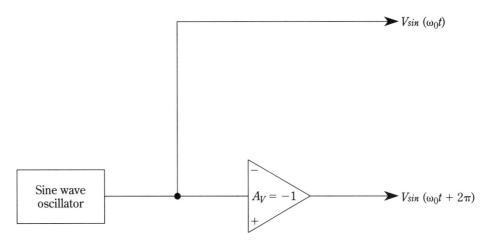

14-9 Biphasic oscillator.

15

Voltage-controlled oscillators

THERE IS A SIGNIFICANT NEED FOR A CIRCUIT THAT WILL PRODUCE A FREQUENCY output that is a function of an applied control voltage. These circuits effectively produce a frequency-modulated output signal. They are used in analog-to-digital conversion, in frequency synthesizers, in instrumentation (especially for data storage or transmission), and in tone signaling and other communications applications. The voltage-controlled oscillator (VCO) and the voltage-to-frequency converter (V/F) circuits fill that need. In this chapter we will take a look at both forms of circuit. They are used somewhat differently, but the end result is essentially the same—an output frequency that is a function of an applied control voltage.

Voltage-controlled oscillators (VCO)—RF

One method for frequency-modulating a radio-frequency (RF) signal is to use a voltage-controlled oscillator (VCO) circuit. Perhaps the most common form of RF VCO is the varactor-tuned RF oscillator. Several different forms of variable RF oscillator are known (see chapter 16), and these are typically called *variable-frequency oscillators* (VFO). It is possible to use most of the VFO designs (Colpitts, Clapp, Hartley, Armstrong, etc.) to make a VCO, although we will select only one or two to demonstrate the principles in this chapter.

VFO circuits are typically tuned by an inductor and a capacitor in an LC network (either series or parallel resonant). If either capacitance or inductance is varied, then the frequency changes a proportional amount. In most cases, it will be the capacitance that is varied, because it is relatively easy to achieve in common circuits with easy-to-obtain components.

The actual amount of frequency change is equal to the square root of the capacitance change, so it takes a fairly considerable capacitance variation to produce a large percentage frequency shift:

$$\frac{\Delta F}{F} = \sqrt{\frac{C}{\Delta C}} \qquad \textbf{(15-1)}$$

Note that an increase in capacitance value causes a decrease in the resonant frequency.

An electronic component called a *varactor* is a voltage-variable-capacitance diode. It is a pn-junction diode that has a junction capacitance proportional to the applied reverse-bias potential. Figure 15-1A shows the usual schematic-diagram circuit symbol for varactor diodes, while Fig. 15-1B shows a diagram of the structure. Like all pn diodes, n-type and p-type semiconductor materials are joined together in a junction; negatively charged carriers (electrons) are present in the n-type material, while positively charged carriers (holes) are present in the p-type material. When the junction is reverse biased, the positive and negative charge carriers are attracted away from the junction, forming a depletion zone that is nearly devoid of holes and electrons. The width of the depletion zone is proportional to the applied reverse-bias potential. A depletion zone is essentially an electrical insulator.

Now consider the situation: two electrically charged materials facing each other across an insulator (i.e., the depletion zone). This structure is the same as for a capacitor, but with the thickness of the dielectric region being proportional to the applied voltage. Thus, the pn-junction capacitance becomes a voltage-variable capacitor. Varactor diodes are especially built to enhance this effect, although it exists to some degree in all pn-junction diodes.

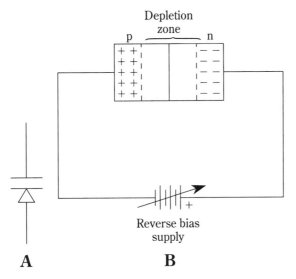

15-1 A. Circuit symbol for a varactor diode. B. Varactor diode structure.

Figure 15-2 shows a Hartley VFO connected as a VCO. The active device is a junction field-effect transistor (JFET), although nearly any IC or transistor device that provides gain at the desired frequency will suffice. The particular device selected is the MPF-102, or its service-grade replacements (e.g., ECG-312 or NTE-

312). The drain terminal is set to ground potential for AC/RF signals because of capacitor C_5, while remaining at a nonzero positive dc potential. A separate dc voltage regulator is supplied to ensure that frequency variations are due only to changes in the applied control voltage, and not due to changes in the applied operating potential.

The resonant circuit consists of C1, L1, and the junction capacitance of varactor diode D1. Note that the inductor is tapped close to its ground end, and the tap is connected to the source of the transistor. This configuration of L1 identifies the circuit as a Hartley oscillator. Trimmer capacitor C1 is not strictly necessary, but forms part of the total capacitance (it is in parallel with the capacitance of D1). The function of the trimmer capacitor is to set the maximum frequency of oscillation when the junction capacitance is minimum. If this feature is not needed, or if it can be accommodated by setting a fixed dc level for the applied control voltage, then the trimmer can be eliminated.

The operating frequency of this circuit is a function of the applied control voltage, V_t. When V_t varies, so will the output frequency. If a sine wave voltage is used, then the circuit will produce sinusoidally swept frequency modulation, but if a sawtooth or triangle wave is applied, then the output frequency is said to be "swept." A square wave applied to the control terminal will produce a "frequency shift keyed" waveform such as used in radioteletype-transmitter operations.

Experiment 15-1

Build the circuit of Fig. 15-2 on perf board, using a 60-pF trimmer for C1, and an ECG-618 or NTE-618 (or equivalent) varactor diode. The inductor should have an in-

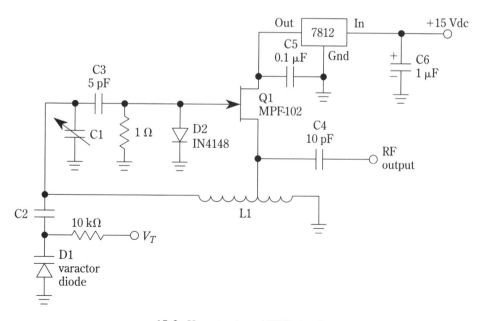

15-2 Varactor-tuned VFO circuit.

ductance around 4 µH. If you opt to use a toroidal-core inductor, then select an Amidon T-50-6 core, and wind it with 20 turns of #26 wire. Connect the output of the oscillator to both an oscilloscope and an HF digital frequency counter. The tuning voltage shall be variable from 0 to 12 V. (If you use another varactor, then set the control-voltage range to be compatible. Most other varactors will want to see 0 to 30 V or so.)

1. Turn on the circuit and observe the operation on the oscilloscope. Synchronize the oscilloscope to show about eight to 12 cycles of the waveform. Vary the control voltage and observe the change of frequency.
2. Adjust the control voltage from 0 to 12 V in steps of 0.5 V. Measure and record on a sheet of paper the frequency of operation on the digital frequency counter at each 0.5-V step.
3. Chart the results of step 2 on a voltage-vs-frequency graph.
4. Apply a 2-Hz sine wave signal to the control input, rather than a dc level. Observe the effects on the oscilloscope. Similarly, apply square waves and triangle waves and observe.

You will notice, if you perform the third step of the experiment, that the frequency is not a linear function of applied voltage. Two factors conspire to prevent that curve from being a straight line. One is the fact that the resonant frequency is proportional to the square root of the product of capacitance and inductance. The second factor is that the junction capacitance is an exponential function of the applied reverse-bias control voltage. If a linear relationship is not needed, then no additional concern is warranted. However, where a more linear relationship is needed, then select a portion of the curve that is nearly linear and avoid the severely curved portions.

Figure 15-3 shows the circuit for a two-varactor UHF voltage-controlled oscillator that is more linear than a single-varactor design. This circuit is sometimes used in frequency-synthesizer applications in radio transmitters and RF instruments. A UHF JFET transistor (2N4416 or equivalent) is used as the active device. Unlike the previous circuit, this circuit places the JFET in the common-gate configuration by grounding the gate terminal for RF through a 470-pF capacitor. By experimenting with values of L_1 less than 0.15 µH, you will find this circuit oscillating in the neighborhood of 300 to 400 MHz, unless stray capacitances from improper construction force it lower.

The type of V/F converter shown in Fig. 15-4A is often used for instrumentation purposes. The actual circuit is shown in Fig. 15-4A, while the timing waveforms are shown in Fig. 15-4B. The operation of this circuit is dependent upon the charging of a capacitor, although not an RC network as in the case of some other oscillator or timer circuits. The input voltage signal (V_x) is amplified (if necessary) by A1, and then converted to a proportional current level in a voltage-to-current converter stage. If the voltage applied to the input remains constant, so will the current output of the V-to-I converter (I).

The current from the V-to-I converter charges the timing capacitor, C. The voltage appearing across this capacitor (V_c) varies with time as the capacitor charges (see the V_c waveform in Fig. 15-4B). The precision discharge circuit is designed to

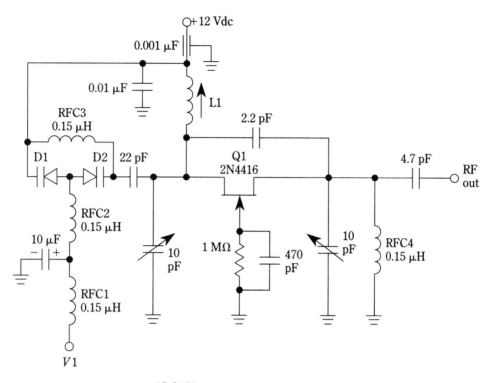

15-3 Varactor-tuned UHF oscillator.

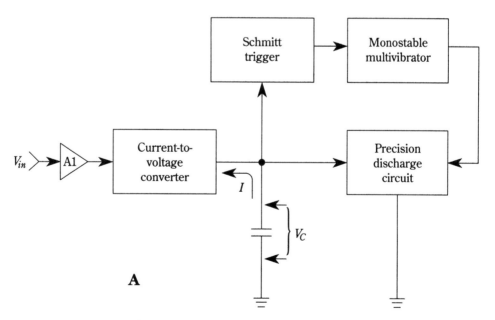

15-4A Voltage-to-frequency converter block diagram.

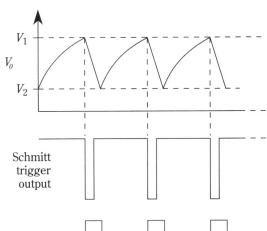

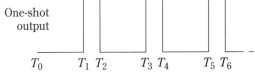

B

15-4B Timing diagram.

discharge capacitor C to a certain level (V_2) whenever the voltage across the capacitor reaches a predetermined value (V_1). When the voltage across the capacitor reaches V_2, a Schmitt-trigger circuit is fired that turns on the precision discharge circuit. The precision discharge circuit, in turn, will cause the capacitor to discharge rapidly, but in a controlled manner, to value V_1. The output pulse snaps high when the Schmitt trigger fires (i.e., the instant V_c reaches V_1) and drops low again when the value of V_c has discharged to V_2. The result is a train of output pulses with a repetition rate exactly dependent upon the capacitor charging current, which, in turn, is dependent upon the applied voltage. Hence, the circuit is a voltage-to-frequency converter.

The NE-566 VCO device

One of the older, but nonetheless still popular, voltage-controlled oscillator circuits is the Signetics 566 device. This chip is widely available at low cost. The NE-566 is described as a voltage-controlled "function generator" in the data books and literature because it produces two different output waveforms: triangle wave (pin 4) and square wave (pin 3). The NE-566 can be programmed for a 10:1 frequency range with proper selection of components. While newer IC devices of a similar vein allow a wide frequency range, the 566 retains the low-cost attribute.

The operating frequency of the NE-566 is set by resistor R1 and capacitor C1, and is found by:

$$f_o = \frac{2\,[(V+) - (V_c)]}{R_1 C_1\,(V+)} \tag{15-2}$$

Where: $V+$ is the supply voltage
V_c is the control voltage applied to pin 5
R_1 is the resistance of R_1 in ohms
C_1 is the capacitance of C_1 in farads

Normally, $2\ k\Omega \leq R_1 \leq 20\ k\Omega$, while the operating frequency can be as high as 1 MHz (1000 kHz).

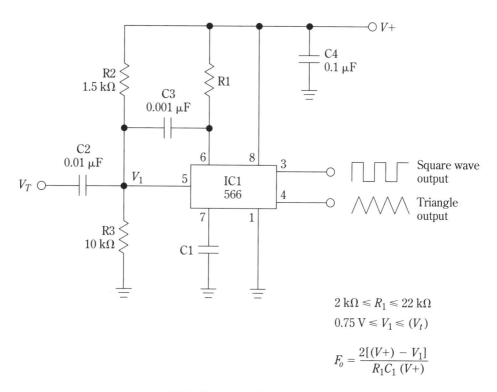

$$2\ k\Omega \leq R_1 \leq 22\ k\Omega$$
$$0.75\ V \leq V_1 \leq (V_t)$$

$$F_o = \frac{2[(V+) - V_1]}{R_1 C_1\ (V+)}$$

15-5 Signetics 566 VCO circuit.

MC-4024P dual VCO chip

The MC-4024P is not a CMOS device, despite the "4xxx" part number. Instead, it is a Motorola TTL-compatible dual voltage-controlled oscillator chip—so do not confuse it with the CMOS 4024 device.

Figure 15-6A shows the internal circuitry of the MC-4024P in block-diagram form. It contains two voltage-controlled oscillators, designated here as OSC A and OSC B. These circuits will oscillate at frequencies up to 25 MHz when a suitable value of timing capacitor (C_A and C_B) are selected. Note that the chip has three different dc power-supply and ground connections, arranged into two different types. The package power (pin 14) and package ground (pin 7) are for the entire MC-4024P device, and must be connected regardless of which oscillator is used. The other dc power and ground terminals are for specific oscillators and are used only when that oscillator is enabled:

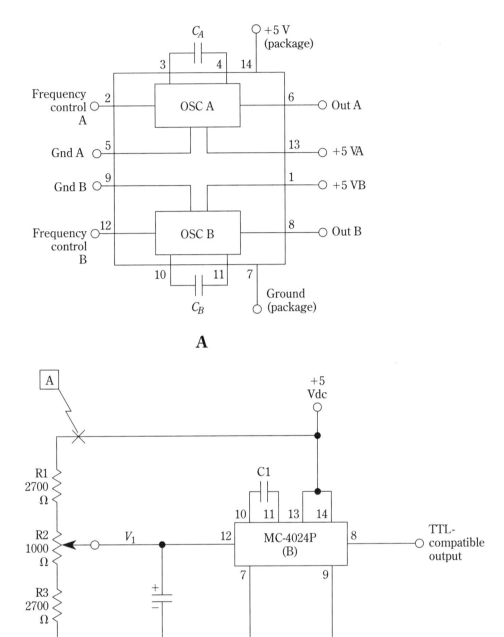

15-6 A. MC-4024P internal circuit. B. practical circuit.

	+5 Vdc Power	**Ground**
OSC A	13	5
OSC B	1	9

All three dc power-supply terminals are designed for +5-Vdc power, as required by the TTL compatibility of the device.

The operating frequency is a function of both the input voltage applied to pin 12 (2 to 5 V) and the value of the timing capacitor. This capacitance is found experimentally, but is approximately:

$$C_{pF} = \frac{300}{f_{\text{MHz}}} \tag{15-3}$$

Figure 15-6B shows the circuit for a typical MC-4024P voltage-controlled oscillator circuit. The control voltage in this case is supplied by a potentiometer voltage divider from the +5-Vdc power supply. However, if the circuit is broken at point "A," use an external signal source to drive the control-voltage input terminal.

16
LC RF variable-frequency oscillators (VFOs)

VARIABLE-FREQUENCY OSCILLATORS (VFOs) ARE RADIO-FREQUENCY SIGNAL generatores that can be continuously tuned using an inductor connected to either an air-variable capacitor, a mica "trimmer" capacitor, or a voltage-tuned variable-capacitance diode (varactor). The VFO differs from the crystal oscillators used in many transmitters in that the frequency can be varied in the VFO, while in the crystal oscillator the frequency is either fixed or variable over only a tiny region. In this chapter, you will learn about VFO circuits by building such circuits as projects.

VFO circuits can be used as signal generators in test equipment, to control ham transmitters, as the local oscillator in either superheterodyne or direct-conversion receiver projects, or in any other applications where a continuously variable source of RF energy is needed. In this chapter are some practical circuits based on easily obtained components, as well as some general guidelines for using and modifying the circuits for your own use.

RF oscillator basics

Both VFOs and crystal oscillators are part of a class of circuits called *feedback oscillators*. Figure 16-1 shows the basic configuration of this type of circuit; it consists of an amplifier with open-loop gain A_{vol}, and a feedback network (which is usually frequency selective) with a "gain" of β. We discussed this subject briefly in chapter 7, so this section is a review.

Barkhausen's criteria

If two conditions are met, then the circuit will oscillate. These criteria are called *Barkhausen's criteria*: 1) the loop gain is unity or greater, and 2) the feedback signal arriving back at the input is phase-shifted 360°. For most practical circuits, with the 180° provided by an inverting amplifier, there must be an additional 180° of phase shift provided by the feedback network. The nature of an inductor-capacitor (LC) tuned circuit is that it can provide this 180° phase shift at only one frequency.

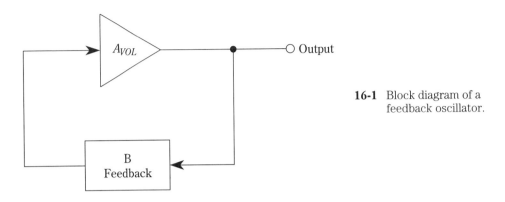

16-1 Block diagram of a feedback oscillator.

Basic categories of VFO circuit

Three general categories of feedback RF oscillator will be considered here: Armstrong oscillators, Hartley oscillators, and Colpitts oscillators (also, a subset of the Colpitts oscillator is the Clapp oscillator). These basic configurations are shown in Fig. 16-2.

The circuit in Fig. 16-2A is the Armstrong oscillator. It is identified by the feedback link positioned as a secondary winding on the tuning coil.

The circuit in Fig. 16-2B is the Hartley oscillator. It is identified by the fact that the resonant-feedback network contains a tapped inductor (which effectively forms an inductive voltage divider).

The Colpitts and Clapp circuits (Fig. 16-2C) are identified by the fact that the feedback network contains a tapped capacitance voltage divider. In both cases, the feedback voltage divider is part of the resonant LC tuning network.

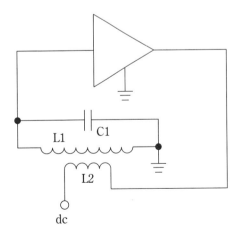

16-2A Armstrong oscillator block diagram.

A

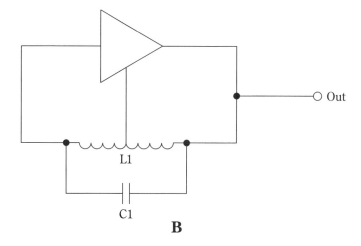

16-2B Hartley oscillator block diagram.

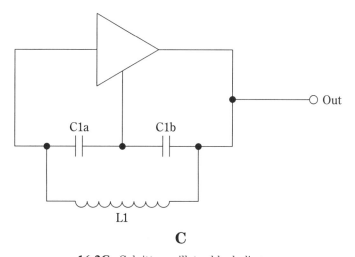

16-2C Colpitts oscillator block diagram.

Power supplies for VFO circuits

It is always a good idea to use only regulated dc power supplies with VFO (or any oscillator) circuits. In fact, most experts agree that a single regulator serving the oscillator is best, because it is not affected by load changes in other circuits on the same side of the regulator. The reason for using only regulated dc is that most oscillators shift frequency a slight amount when the power supply voltages change. If you are a ham radio operator familiar with CW (i.e., "Morse code") transmission, then you will recognize such variation as the tranmission defect called "chirp." Although not all of the oscillator circuits in this chapter show the regulator, it is a good idea to use one anyway.

Some basic circuit configurations

The basic circuits discussed above can be configured in any of several different ways. In this chapter, we will use one of three active devices as the amplifier portion: junction field-effect transistor (JFET), metal-oxide field-effect transistor (MOSFET), and the Signetics NE-602 RF-balanced-mixer integrated circuit. The JFET and MOSFET parts were selected to be easily available.

The basic JFET is the MPF-102 device, while the MOSFET is the 40673 device. If you buy parts from a distributor who carries a radio-TV service replacement line, such as ECG or NTE, then you will find these parts easily available in those lines as well. The MPF-102 is replaced by the ECG-312 and the NTE-312, while the 40673 is replaced by the ECG-222 and NTE-222 devices. Bipolar transistors can also be used, but the JFET and MOSFET circuits are very well behaved, and will suffice for most applications in practical projects.

Input-side circuit configurations

Figure 16-3 shows two input configurations, one each for MPF-102 and 40673. In Fig. 16-3A is the circuit for the JFET device. It consists of a gate resistor of 100 kΩ to ground, and a diode (1N914, 1N4148, or equivalent). In many cases, you will need to use a capacitor in the gate circuit (C1), especially if there is a dc source or ground directly in the circuit. To prevent loading of the tuned circuit, it is customary to make the coupling capacitor small compared to the tuning capacitor; typically, values from 2 to 10 pF are used for HF and MW VFO circuits.

The same circuit can be used on the MOSFET, but a dc bias circuit will be needed for the second gate, G2. The bias network shown in Fig. 16-3B (R_2/R_3) is set to bias G2 to about ⅓ the $V+$ voltage. I've used equal value resistors (10 kΩ and 10 kΩ) for R_2/R_3. A bypass/decoupling capacitor (C2) sets G2 to a low impedance for RF, while keeping it at the bias voltage for dc.

The diode in the input circuit perplexes some people when they first see it in oscillator circuits. The function of the diode is to clean up the signal, and make it closer

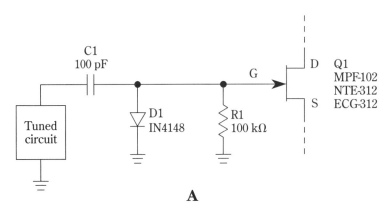

16-3A Gate circuit for JFET oscillators.

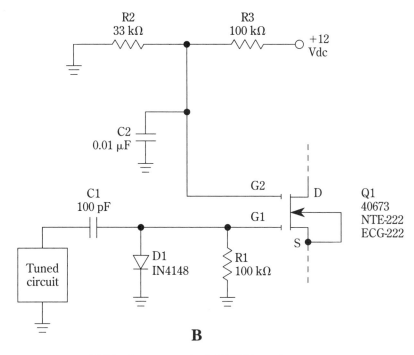

B

16-3B Gate circuit for MOSFET oscillators.

to a low harmonic sine wave (all nonsine-wave or distorted sine waves have harmonic content—by definition a "pure" sine wave has no harmonics). Figure 16-4A shows the sine-wave output signal when the diode is connected. Note that the waveform is a reasonably good sine wave. The waveform in Fig. 16-4B is a distorted sine wave, and has a higher amplitude than the case of Fig. 16-4A.

The tuned circuit can be either a parallel tuned circuit (Fig. 16-5A), in the case of Colpitts or Hartley oscillators, or series tuned circuit (Fig. 16-5B), in the case of Clapp oscillators. In the case of the Hartley oscillator, the inductor (L1) will be tapped. In any LC resonant circuit, resonance is that point where the inductive reactance (X_L) and capacitive reactance (X_C) are equal to each other. Because these elements cancel out, the impedance of such a circuit is resistive.

It is generally true that there should be a high C/L ratio, so it is common practice to select a relatively small inductance and match it with a higher capacitance. Many of the oscillators in this chapter are designed for the middle of the 1 to 10-MHz band, and use inductances on the order of 3.3 to 7 μH. These inductances are relatively easy to obtain when using either "solenoid" (cylindrical) or "toroidal" coil forms. See the Amidon Associates[1] catalog for information on winding both sorts of coils. Also, Barker & Williamson[2] makes air-core coil stock that is suitable for most HF oscillators. Both fixed-value and variable inductors can be used for these circuits.

[1]P.O. Box 956, Torrance, CA 90508.
[2]10 Canal Street, Bristol, PA, 19007; 215-788-5581.

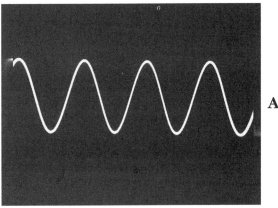

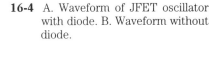

16-4 A. Waveform of JFET oscillator with diode. B. Waveform without diode.

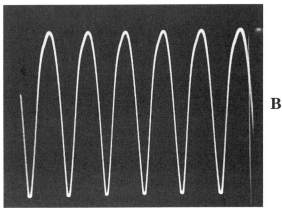

The capacitance can be made up from more than one capacitor, and this is generally the best practice. The change of frequency is proportional to the square root of the ratio of the capacitance change:

$$\frac{f_{max}}{f_{min}} = \frac{C_{max}}{C_{min}} \qquad\qquad \textbf{(16-1)}$$

That is why a variable capacitor for the AM broadcast band consists of a 25-pF to 365-pF air-variable capacitor shunted by a 25-pF (or so) trimmer capacitor (which is usually set to around 15 pF). The 3.08:1 ratio of maximum to minimum capacitance is more than sufficient to cover the 3.02:1 ratio of the maximum and minimum frequencies of the AM broadcast band.

Selecting values of L and C is a somewhat tedious and iterative affair. You are advised to sit down with a calculator and make a few trials. Part of the problem comes from the fact that both fixed and variable capacitors come in standard values that may or may not be exactly what's needed. Juggle the inductance (which is easy to wind to a custom value) to provide the desired frequency change with the available variable capacitors.

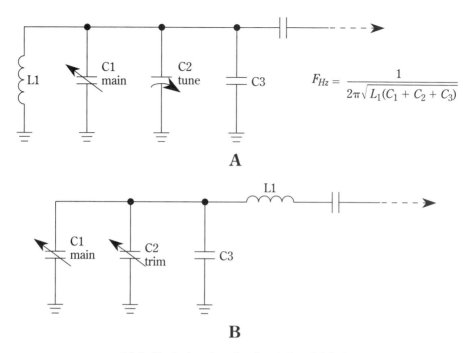

$$F_{Hz} = \frac{1}{2\pi\sqrt{L_1(C_1 + C_2 + C_3)}}$$

16-5 Typical tuning circuits: A. Parallel B. Series.

It is common practice to make the total of the fixed capacitors plus the maximum values of the variable (main tuning and trimmer) capacitors somewhat larger than the total required to resonate at the very lowest frequency in the desired range. When the trimmer is set to a value less than maximum, the total capacitance will be close to the desired value.

Project 16-1 Hartley JFET VFO

The Hartley oscillator, as previously described, is identified by a feedback path that includes a tapped inductor; the inductor is also part of the resonant tuning circuit of the oscillator. Figure 16-6 shows a simple Hartley oscillator based on the MPF-102 junction field-effect transistor (JFET). The output signal is taken through a small-value capacitor (to limit loading) connected to the emitter of the transistor.

The frequency of oscillation is set by the combined effect of L_1, C_1, and C_2:

$$f_{Hz} = \frac{1}{2\pi\sqrt{(L_1)(C_1 + C_2)}} \tag{16-2}$$

To resonate at 5-MHz with a 5-μH inductor, a total capacitance of about 200 pF is required. Because of stray capacitances and errors in the values of actual capacitors, it is common practice to use more total capacitance than needed, and use variable capacitors to trim. For example, we could use a 140-pF variable capacitor for the main tuning, and an 80-pF trimmer to set the maximum to the required value.

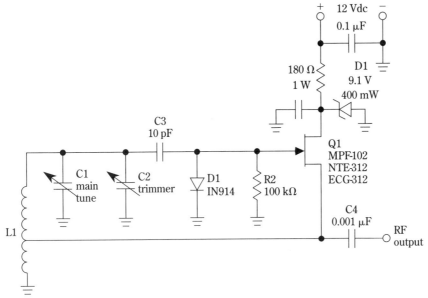

16-6 Simple Hartley JFET oscillator.

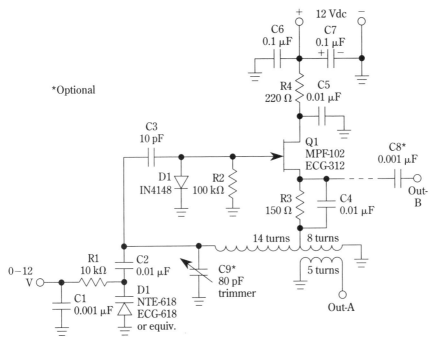

16-7 Voltage-tuned JFET Hartley oscillator.

Figure 16-7 shows a 5-MHz Hartley VFO circuit based on the MPF-102 JFET transistor. It is very similar to the previous circuit in basic concept, but there are some differences. The most significant difference is the use of a variable capacitance diode (varactor) instead of the main tuning capacitor. The diode shown here is a 14- pF to 440-pF varactor used to tune the AM broadcast band in radio receivers. Another significant difference is that the output signal is taken from a secondary winding on the tuning inductor. This winding consists of fewer turns than the lower portion of the main tuning inductor. Care must be taken not to load down this circuit by connecting a varying load resistance across the output winding.

The circuit of Fig. 16-7 can be made to sweep the entire frequency range by applying a sawtooth waveform that rises from 0 V to +12 V. This type of circuit makes it relatively easy to build a sweep-generator circuit or a swept-tuned radio receiver. In general, the sweep rate should be around 40 Hz if the detector has a narrowband filter. Other sweep frequencies are usable as well, but care must be taken not to "ring" any following resonant circuits or filters because of a too-fast sweep rate.

The tuning properties of a varactor LC circuit are nonlinear because of the nature of varactor diodes. A graph of tuning voltage versus frequency should be made, and only the linear portion of the curve should be used for sweep purposes.

The circuit in Fig. 16-8A is capable of producing as much as several volts of signal in the 1 to 10-MHz range. The actual tuning range depends on the particular components used. The heart of the oscillator is an MPF-102 JFET (Q1). Two different tuning schemes are provided. For a limited range, the main tuning is provided by a 365-pF AM-broadcast-band tuning capacitor, and the tuning range is 5000 to 5500 kHz. In that case, disconnect and do not use the varactor diode (D2). The alternate scheme deletes $C1$, and uses either a 365-pF variable or the varactor circuit (shown) at point "A." This version has a rather wider tuning range for the same capacitance change. In the previous circuit, the tuning range was reduced by the capacitor divider action of C4 and C5.

The circuit of Fig. 16-8A can be modified by adding or subtracting capacitors from the tuning network. As shown, C1, C2, C3, C4, and $C5$ are all part of the tuning circuit. The 126-pF fixed capacitor (needed for 5 to 5.5-MHz operation) is made up from parallel 82-pF and 47-pF capacitors. A version of the circuit using the NTE-618 or ECG-618 440-pF diode was built without the mica trimmer capacitor (C2). It produced a voltage (V_t)-versus-frequency characteristic shown in Fig. 16-8B (this curve is for a circuit with C2 removed). The varactor is comfortable over a range of 0 to +12 V (although my dc power supply only goes down to +1.26 V, which is why the offset at 2.2 MHz). The tuning range is 2.2 MHz to 4.1 MHz, but can be lowered by either increasing any of the capacitors (e.g., C3) or eliminating or reducing the capacitors (C1 – C5).

The main inductor (L1) consists of 35 turns of #28 enameled wire on an Amidon T-50-2 (red) toroidal core. Place the tap at eight turns from the ground end. Form the tap by winding two separate, but contiguous, windings, one of eight turns and one of 27 turns. Connect these two windings together electrically at the point where they come together. Solder together the two wire ends of the two coils and use as a single wire to connect to the source ("S") of the JFET oscillator.

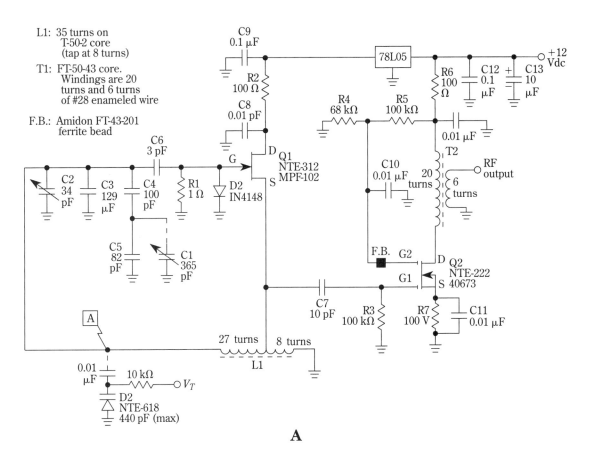

L1: 35 turns on T-50-2 core (tap at 8 turns)

T1: FT-50-43 core. Windings are 20 turns and 6 turns of #28 enameled wire

F.B.: Amidon FT-43-201 ferrite bead

A

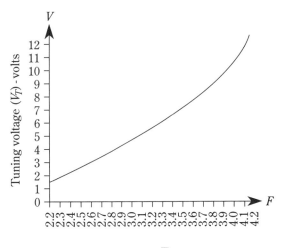

B

16-8 A. Hartley oscillator with buffer amplifier. B. Tuning characteristic.

The drain ("D") of the JFET is kept at a ground potential for RF signals by the bypass capacitor, C_8. The output signal is taken from the source ("S") terminal through a 10-pF capacitor. This signal is fed to gate 1 ("G1") of the 40673 dual-gate MOSFET that is used as an output buffer-amplifier stage. This circuit produces a 44-dB gain, and is responsible for making the signal voltage so large. The output transformer (T1) consists of an Amidon Associates FT-50-43 core wound with 20 turns of #28 enameled wire for the primary, and six turns of the same wire for the secondary.

The output signal of Fig. 16-8A is huge as RF oscillators go, so it may have to be attenuated in some cases. Connect an attenuator resistor pad in series with the output signal, or reduce the dc bias voltage on the MOSFET gate 2. Reduce the value of R_4 until the desired output signal is achieved (do not reduce it below about 1 kΩ).

Project 16-2 Clapp VFO circuit

The Colpitts and Clapp oscillators are very similar to each other in that both depend on a capacitor voltage divider (C4 and C5 in Fig. 16-9) for feedback. The difference in the two oscillator circuits is that a Colpitts oscillator uses a parallel-resonant LC circuit, while a Clapp oscillator uses a series-resonant LC circuit. The circuit in Fig. 16-9 is a Clapp oscillator because of the series-tuned LC network. This circuit can be used from 0.5 to 7 MHz, and even higher if C_4 and C_5 are reduced to about 100 pF.

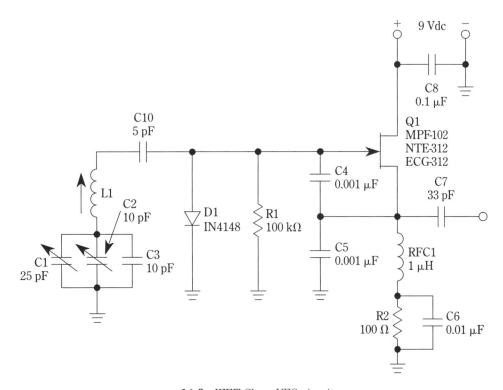

16-9 JFET Clapp VFO circuit.

The output from this circuit is taken from the source ("S") of the JFET oscillator transistor. The output circuit includes an RF choke (RFC1) that builds the output amplitude. Bias voltage for the JFET is provided by R_2 and the source-drain current flowing through it.

Project 16-3 NE-602 VFO circuits

The Signetics NE-602AN device is a double-balanced modulator (DBM) and oscillator integrated circuit. Normally, it is used as the RF "front-end" of radio receivers, but if the DBM is unbalanced by placing a 10-kΩ resistor from the RF input (pin no. 1) to ground, it will function as an oscillator that produces about 500-mV output signal.

Figure 16-10 shows an NE-602AN Colpitts oscillator circuit. Three capacitors (C1, C2, and C3) are used in this circuit, rather than two, because of a need for dc blocking. These capacitors should be equal to each other, and have a value on the order of 2400 pF/f_{MHz}. The inductor should have an approximate value of 7

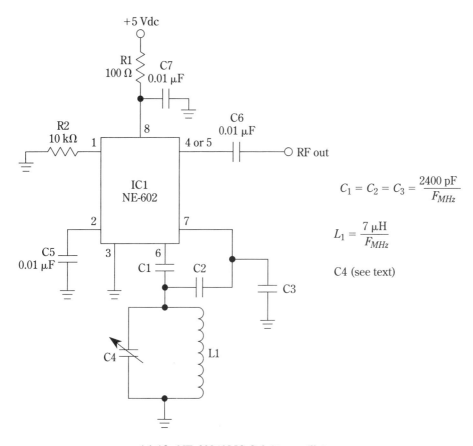

$$C_1 = C_2 = C_3 = \frac{2400 \text{ pF}}{F_{MHz}}$$

$$L_1 = \frac{7 \text{ μH}}{F_{MHz}}$$

C4 (see text)

16-10 NE-602AN IC Colpitts oscillator.

$\mu H/f_{MHz}$. The tuning capacitor, C4, should have a value that will resonate with the selected inductor:

$$C_4 = \frac{1}{4\pi^2 f^2 L_1} \tag{16-3}$$

For example, a 5000-kHz (5-MHz) oscillator should have network capacitors of 2400 pF/5 MHz, or 480 pF (use 470-pF standard value capacitors). The inductor should be 1.4 µH (17 turns on an Amidon T-50-2 (red) core). To resonate with the 1.4-µH inductor requires 723 pF. However, 470 pF/2, or 236 pF, are already in the circuit because of the series network C_2/C_3. Thus, a variable capacitor (C4) of 723 pF –236 pF, or 487 pF, is used.

An NE-603AN Hartley oscillator is shown in Fig. 16-11. This circuit is identified by the tapped coil in the LC network. The value of the inductor is about 10 $\mu H/f_{MHz}$, and is tapped from one-fourth to one-third the way from the ground end. The capacitor needs to resonate at the desired frequency. For our 5-MHz example, use an inductor of 2 µH, which means 20 turns of wire on a T-50-2 (red) core.

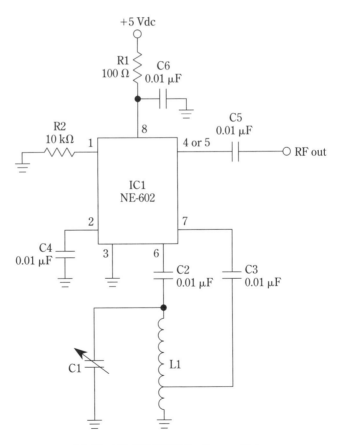

16-11 NE-602AN IC Hartley oscillator.

A voltage-tuned Clapp NE-602AN oscillator is shown in Fig. 16-12. This circuit uses a varactor diode to set the operating frequency. With the 100-pF capacitors shown, this circuit has oscillated from about 6 MHz to about 15 MHz, using an NTE-614 (33 pF) diode.

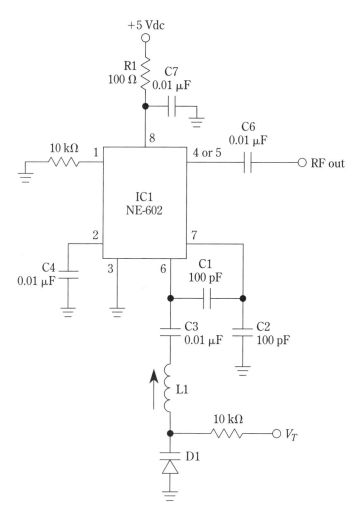

16-12 Voltage-tuned NE-602AN IC Clapp oscillator.

17
The 555 integrated circuit timer device

IN CHAPTER 11, YOU LEARNED ABOUT MONOSTABLE AND ASTABLE MULTIVIBRATOR circuits. These circuits produce square output signals, and were based on the operational amplifier. The circuits in this chapter are also monostable or astable multivibrators, but are based on the 555 timer device.

The 555 integrated circuit (IC) timer represents a class of chips that are extraordinarily well-behaved, and very easy to apply. These timers are based on the properties of the series RC timing network (see chapter 9) and the voltage comparator. A combination of voltage-comparator circuits and digital circuits are used inside the 555 chips.

Although several devices are on the market, the most common and best known is the type 555 device. The 555 is now made by a number of different semiconductor manufacturers, but was originated by Signetics in 1970. Today, the 555 remains one of the most widespread, best-selling IC devices on the market, rivaling even some general-purpose operational amplifiers in numbers sold.

The original Signetics products included the SE-555, which operated a temperature range of −55 to +125°C, and the NE-555, which operated over the range 0 to +70°C. Several different designations are now commonly used for the 555 made by other makers, including simply "555" and "LM-555" or some variant of these. A dual 555-class timer is also marketed under the number "556." There is also a low-power CMOS version of the 555 marketed as the LMC-555.

The 555 is a multipurpose chip that operates at dc power-supply potentials from +5 Vdc to +18 Vdc. The temperature stability of these devices is on the order of 50 PPM/°C (i.e., 0.005%/°C). The output of the 555 can either sink or source up to 200 mA of current. It is compatible with TTL devices (when the 555 is operated from a +5-Vdc power supply), CMOS devices, operational amplifiers, other linear IC devices, transistors, and most classes of solid-state devices. The 555 will also operate with most passive electronic components.

Several factors contribute to the popularity of the 555 device. Besides the versatile nature of the device, it is well-behaved in the sense that operation is straightforward and circuit designs are generally simple. Like the general-purpose operational amplifier, the 555 usually works in a predictable manner, according to the standard published equations.

The 555 operates in two different modes: monostable (one-shot) and astable (free-running). Figure 17-1A shows the astable-mode output from pin 3 of the 555. The waveform is a series of square waves that can be varied in duty cycle over the range 50 to 99.9 percent, and in frequency from less than 0.1 Hz to more than 100 kHz. Monostable operation (Fig. 17-1B) requires a trigger pulse applied to pin 2 of the 555. The trigger must drop from a level $>2(V+)/3$ down to $<(V+)/3$. Output pulse durations from microseconds up to hours are possible. The principal constraint on longer operation is the leakage resistance of the capacitor used in the external timing circuit.

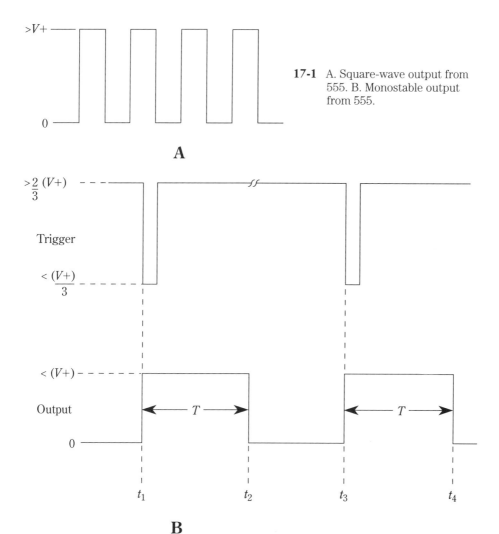

17-1 A. Square-wave output from 555. B. Monostable output from 555.

Pinouts and internal circuits of the 555 IC timer

The package for the 555 device is shown in Fig. 17-2. Most 555s are sold in the eight-pin miniDIP package as shown, although some are found in the eight-pin metal-can IC package. The latter are mostly the military-specification temperature-range SE-555 series. The pinouts are the same on both miniDIP and metal-can versions.

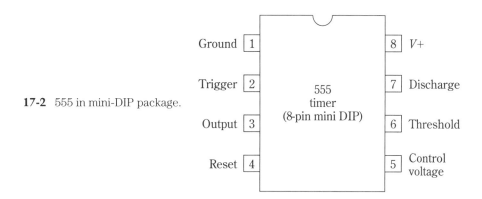

17-2 555 in mini-DIP package.

The internal circuitry is shown in block form in Fig. 17-3. The following stages are found: two voltage comparators (COMP1 and COMP2), a reset-set (RS) control flip-flop (which can be reset from outside the chip through pin 4), an inverting output amplifier (A1), and a discharge transistor (Q1). The bias levels of the two comparators are determined by a resistor voltage divider (R_a, R_b, and R_c) between $V+$ and ground. The inverting input of COMP1 is set to $2(V+)/3$, and the noninverting input of COMP2 is set to $(V+)/3$.

Figures 17-2 and 17-3 show the pinouts of the 555. They should be memorized. In the descriptions below, the term "high" implies a level $>2(V+)/3$, and "low" implies a grounded condition ($V = 0$), unless otherwise specified in the discussion. These pins serve the following functions:

Ground (pin 1) The ground pin serves as the common reference point for all signals and voltages in the 555 circuit, both internal and external to the chip.

Trigger (pin 2) The trigger pin is normally held at a potential $>2(V+)/3$. In this state, the 555 output (pin 3) is low. If the trigger pin is brought low to a potential $<(V+)/3$, then the output (pin 3) abruptly switches to the high state. The output remains high as long as pin 2 is low, but the output does not necessarily revert back to low immediately after pin 2 is brought high again (see operation of the threshold input, later).

Output (pin 3) The output pin of the 555 is capable of either sinking or sourcing current up to 200 mA. This operation is in contrast to other IC devices in which the outputs of various devices will either sink or source current, but not both. Whether the 555 output operates as a sink or a source depends on the configuration of the external load. Figure 17-4 shows both types of operation.

In Fig. 17-4A, the external load (R_L) is connected between the 555 output and $V+$. Current only flows in the load when pin 3 is low. In that condition, the external

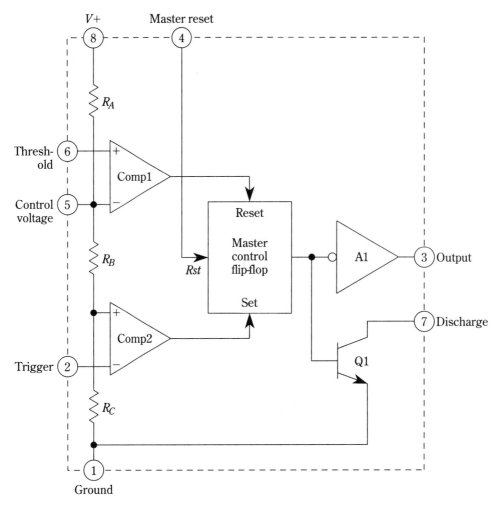

17-3 Internal circuit of 555.

load is grounded through pin 1 and a small internal source resistance, R_{S1}. In this configuration the 555 output is a current sink.

The operation depicted in Fig. 17-4B is applicable when the load is connected between pin 3 of the 555 and ground. When the output is low, the load current is zero. When the output is high, however, the load is connected to $V+$ through a small internal resistance R_{S2} and pin 8. In this configuration the output serves as a current source.

Reset (pin 4) The reset pin is connected to a preset input of the 555 internal control flip-flop. When a low is applied to pin 4, the output of the 555 (pin 3) switches immediately to a low state. In normal operation it is common practice to connect pin 4 to $V+$ to prevent false resets from noise impulses.

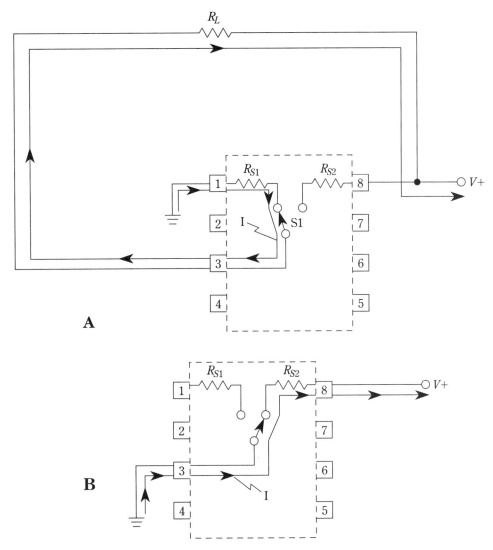

17-4 A. Current sink operation. B. Current source operation.

Control voltage (pin 5) The control-voltage pin normally rests at a potential of $2(V+)/3$ due to an internal resistive voltage divider (see R_a through R_c in Fig. 17-3). Applying an external voltage to this pin, or connecting a resistor to ground, will change the duty cycle of the output signal. If not used, pin 5 should be decoupled to ground through a 0.01-μF to 0.1-μF capacitor.

Threshold (pin 6) The threshold pin is connected to the noninverting input (+IN) of comparator COMP1, and is used to monitor the voltage across the capacitor in the external RC timing network. If pin 6 is at a potential of $<2(V+)/3$, then the output of the control flip-flop is low, and the output (pin 3) is high. Alternatively, when the voltage on pin 6 is $\geq 2(V+)/3$, then the output of COMP1 is high and chip output (pin 3) is low.

Discharge (pin 7)   Connect the discharge pin to the collector of npn transistor Q1, and connect the emitter of Q1 to the ground pin (pin 1). Connect the base of Q1 to the NOT-Q output of the control flip-flop. When the 555 output is high, the NOT-Q output of the control flip-flop is low, so Q1 is turned off. The c-e resistance of Q1 is very high under this condition, so does not appreciably affect the external circuitry. But when the control flip-flop NOT-Q output is high, however, the 555 output is low and Q1 is biased hard on. The c-e path is in saturation, so the c-e resistance is very low. Pin 7 is effectively grounded under this condition.

V+ power supply (pin 8)   Connect the dc power supply between ground (pin 1) and pin 8, with pin 8 being positive. In good practice, a 0.1-μF to 10-μF decoupling capacitor will normally be used between pin 8 and ground.

Monostable operation of the 555 IC timer

A monostable multivibrator (MMV), also called the "one-shot" circuit, produces a single output pulse of fixed duration when triggered by an input pulse. The output of the one-shot will snap high following the trigger pulse, and will remain high for a fixed, predetermined duration. When this time expires, the one-shot is "timed-out," so the output snaps low again. The output of the one-shot will remain low indefinitely unless another trigger pulse is applied to the circuit. The 555 can be operated as a monostable multivibrator by suitable connection of the external circuit.

Figure 17-5 shows the operation of the 555 as a monostable multivibrator. To make the operation of the circuit easier to understand, Fig. 17-5A shows the internal circuitry as well as the external circuitry; Figure 17-5B shows the timing diagram for this circuit (Fig. 17-5C shows the same circuit in the more conventional schematic-diagram format).

The two internal voltage comparators are biased to certain potential levels by a series voltage divider consisting of internal resistors R_a, R_b, and R_c. The inverting input of voltage comparator COMP1 is biased to $2(V+)/3$, while the noninverting input of COMP2 is biased to $(V+)/3$. These levels govern the operation of the 555 device in whichever mode is selected. An external timing network $(R_1 C_1)$ is connected between $V+$ and the noninverting input of COMP1 via pin 6. Also connected to pin 6 is 555 pin 7, which has the effect of connecting the transistor across capacitor C_1. If the transistor is turned on, then the capacitor looks into a very low resistance short circuit through the c-e path of the transistor.

When power is initially applied to the 555, the voltage at the inverting input of COMP1 will go immediately to $2(V+)/3$ and the noninverting input of COMP2 will go to $(V+)/3$. The control flip-flop is in the reset condition, so the NOT-Q output is high. Because this flip-flop is connected to output pin 3 through an inverting amplifier (A1), the output is low at this point. Also, because NOT-Q is high, transistor Q1 is biased into saturation, creating a short circuit to ground across external timing capacitor C_1. The capacitor remains discharged in this condition ($V_c = 0$).

If a trigger pulse is applied to pin 2 of the 555, and if that pulse drops to a voltage that is less than $(V+)/3$, as shown in Fig. 17-5B, then comparator COMP2 sees a situation where the inverting input is less positive than the noninverting input, so the output of COMP2 snaps high. This action sets the control flip-flop, forcing the

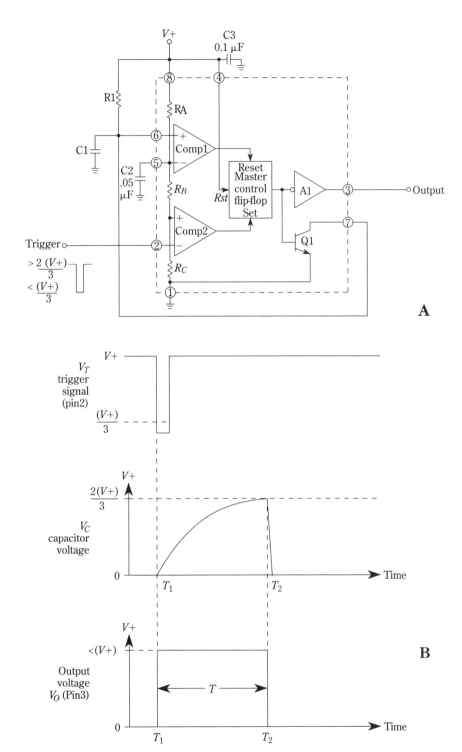

17-5 Monostable multivibrator circuit: A. With internal 555 circuit shown. B. Timing diagram.

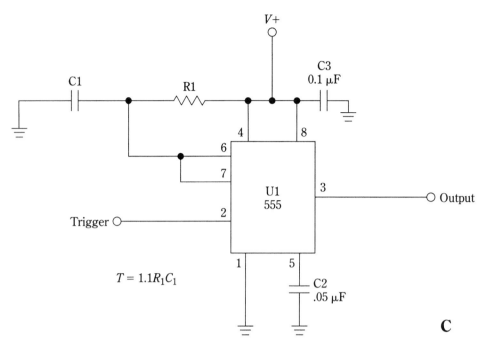

17-5C Schematic form of circuit.

NOT-Q output low, and therefore the 555 output high. The low at the output of the control flip-flop also means that transistor Q1 is now unbiased, so the short across the external capacitor is removed. The voltage across C_1 begins to rise (see Figs. 17-5B and 17-5C). The voltage will continue to rise until it reaches 2($V+$)/3, at which time comparator COMP1 will snap high, causing the flip-flop to reset. When the flip-flop resets, its NOT-Q output drops low again, terminating the output pulse and re-turning the capacitor voltage to zero. The 555 will remain in this state until another trigger pulse is received.

The timing equation for the 555 can be derived in exactly the same manner as the equations used with the operational-amplifier MMV circuits. The basic RC net-work equation was discussed in chapter 16, and relates the time required for a ca-pacitor voltage to rise from a starting point (V_{C1}) to an end point (V_{C2}) with a given RC time constant:

$$T = -RC \ln\left(\frac{V - V_{C2}}{V - V_{C1}}\right) \qquad (17\text{-}1)$$

In the 555 timer the voltage source is $V+$, the starting voltage is zero, and the trip-point voltage for comparator COMP1 is 2($V+$)/3. Equation 17.1 can therefore be rewritten as:

$$T = -R_1 C_1 \ln\left(\frac{V - V_{C2}}{V - V_{C1}}\right) \qquad (17\text{-}2)$$

$$= -R_1C_1\ln\left(\frac{(V+) - 2(V+)/3}{V+}\right)$$

$$= -R_1C_1\ln(1 - 0.667)$$

$$= -R_1C_1\ln(0.333)$$

$$T = 1.1\,R_1C_1 \tag{17-3}$$

Input triggering methods for the 555 MMV circuit

Trigger the 555 MMV circuit by bringing pin 2 from a positive voltage down to a level that is less than $(V+)/3$. Triggering can be accomplished by applying a pulse from an external signal source, or through other means. Figure 17-6 shows the circuit for a simple push-button switch trigger circuit. Connect a pull-up resistor (R2) between pin 2 and $V+$. If normally open push-button switch S1 is open, then the trigger input is held at a potential very close to $V+$. But when S1 is closed, pin 2 is brought low to ground potential. Because pin 2 is now at a potential less than $(V+)/3$, the 555 MMV will trigger. This circuit can be used for contact debouncing.

Experiment 17-1

This experiment examines the operation of the monostable multivibrator ("one-shot") configuration of the 555 IC timer device. The time duration of the output signal is set

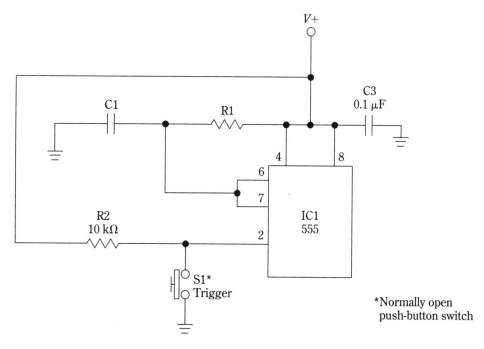

17-6 Push-button triggering of 555.

long enough that you can explore the operation without an oscilloscope. If you use a scope, you can reduce the values to C1 = 0.05 µF and R1A = 33 kΩ, and R1B = 20 kΩ.

The output indicator is a light-emitting diode (LED) in series with a current-limiting resistor (R3, 820 Ω). The 555 device is able to serve as both a current source and a current sink. In this circuit, we use the current-source capability, so the LED and resistor are connected from pin 3 to ground.

1. Build the circuit of Fig. 17-EXP1. Switch S1 is a normally open push-button type that closes when pressed.
2. Turn on the circuit; the LED should be off.
3. Set R1B to the minimum value (R1B = 0 Ω).
4. Calculate the output time duration using the equation:

$$T = 1.1(\text{R1A} + \text{R1B})\,C1$$

5. Connect a high-impedance voltmeter across $C1$.
6. Press S1 while watching the voltmeter and the output LED. The output duration can be measured with a watch or clock equipped with a sweep second hand.

    The LED should turn on for a period of time approximately equal to the duration calculated previously (the difference being due to measurement errors and tolerances of the resistor and capacitor values).
7. Repeat the process using various settings of R1B from minimum to maximum value.

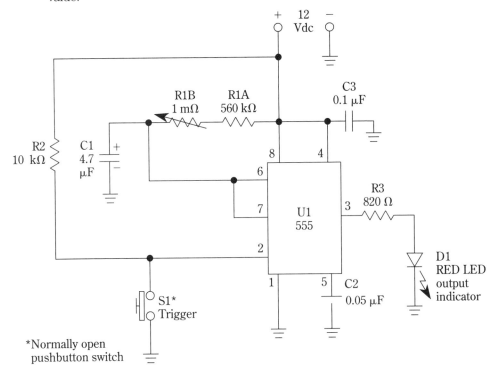

17-EXP1 Circuit for experiment 17-1.

A circuit for inverting the trigger pulse applied to the 555 is shown in Fig. 17-7. In this circuit an npn bipolar transistor is used in the common-emitter mode to invert the pulse. Again, a pull-up resistor is used to keep pin 2 at $V+$ when the transistor is turned off. However, when the positive-polarity trigger pulse is received at the base of transistor Q1, the transistor saturates and forces the collector (and pin 2 of the 555) to near ground potential.

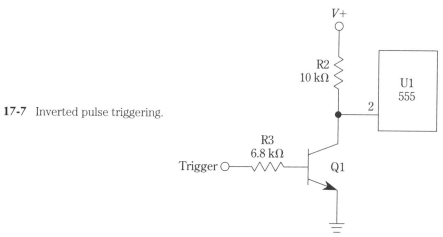

17-7 Inverted pulse triggering.

Figure 17-8 shows two ac-coupled versions of the trigger circuit. In these circuits a pull-up resistor keeps pin 2 normally at $V+$. However, when a pulse is applied to the input end of capacitor C4, a differentiated (i.e., spiked) version of the pulse is created at the trigger input of the 555. Diode D1 clips the positive-going spike to 0.6 or 0.7 V below $V+$, passing only the negative-going pulse to the 555. If the negative-going spike can counteract the positive bias provided by R_2 sufficiently to force the voltage lower than $(V+)/3$, then the 555 will trigger. A push-button switch version of this same circuit is shown in Fig. 17-8B.

A touchplate trigger circuit is shown in Fig. 17-9. The pull-up resistor R2 has a very high value (22 MΩ shown here). The touchplate consists of a pair of closely spaced electrodes. As long as there is no external resistance between the two halves of the touchplate, the trigger input of the 555 remains at $V+$. However, when a resistance is connected across the touchplate, the voltage (V_1) drops to a very low value. If the average finger resistance is about 20 kΩ, the voltage drops to:

$$V_1 = \frac{(V+) \ (20 \ \text{k}\Omega)}{(R_2 + 20 \ \text{k}\Omega)} \tag{17-4}$$

Which, when $R_2 = 22$ MΩ, to $0.0009(V+)$—which is certainly less than $(V+)/3$.

Experiment 17-2

Repeat Experiment 17-1 using a trigger circuit such as Fig. 17-9 instead of the manual triggering circuit (R2/S1) used in Fig. 17-EXP1.

The same concept is used in the liquid level detector shown in Fig. 17-10. Once again, a 22-MΩ pull-up resistor is used to keep pin 2 at $V+$ under normal operation.

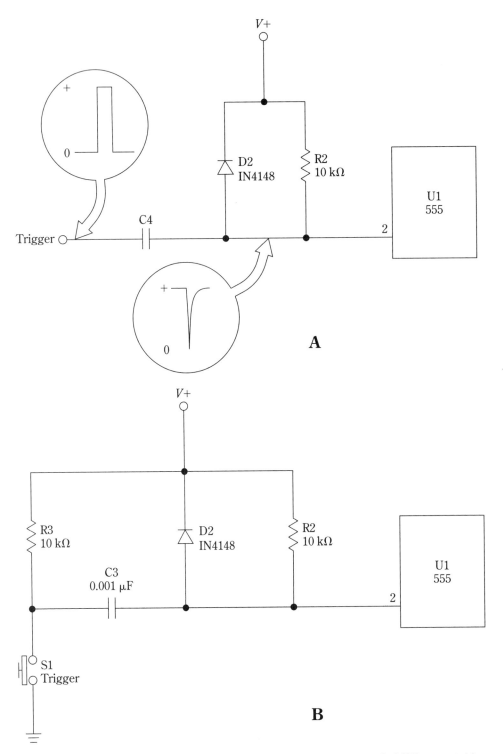

17-8 A. ac-coupled triggering. B. Push-button triggering of the ac-coupled 555 monostable.

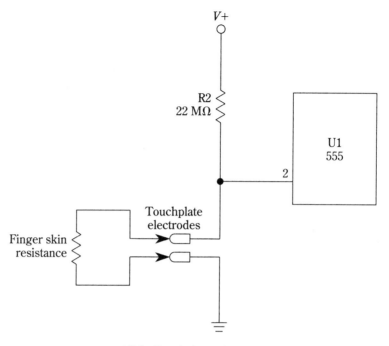

17-9 Touchplate triggering.

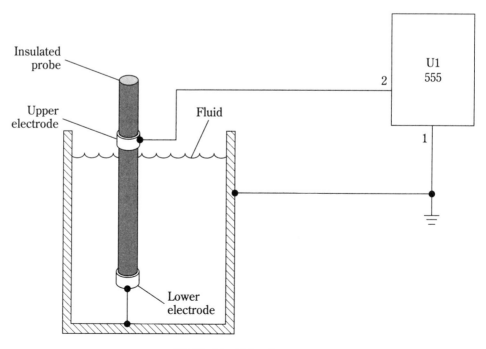

17-10 Liquid-level sensor.

When the liquid level rises sufficiently to short out the electrodes, however, the voltage on pin 2 (V_1) drops to a very low level, forcing the 555 to trigger.

Retriggerable operation of the 555 MMV circuit

The 555 is a nonretriggerable monostable multivibrator. If additional trigger pulses are received prior to the timeout of the output pulse, then the additional pulses have no effect on the output. However, the first pulse after timeout occurs will cause the output again to snap high.

The circuit in Fig. 17-11 will permit retriggering of the 555 device. Keep in mind that the 555 IC timer is normally a standard nonretriggerable monostable multivibrator. In other words, the 555 will not retrigger once triggered, until the timeout of the first pulse. Figure 17-11A shows the operation of a *retriggerable monostable multivibrator*. At time t_1, a trigger pulse is received, so the output snaps high for a period of time T, determined by 1.1RC. This duration would normally expire at t_3. At time t_2, however, a second trigger pulse is received, so the output triggers for another period T. The total period, T', is now T plus the expired portion of the original output pulse. In other words:

$$T'=T + (t_1 - t_2) \qquad\qquad (17\text{-}5)$$

Figure 17-11B shows a trigger-pin circuit that will make a 555 into a retriggerable monostable multivibrator. An external npn transistor (Q2 in Fig. 17-11B) is connected with its c-e path across timing capacitor C1. In this sense it mimics the internal dis-

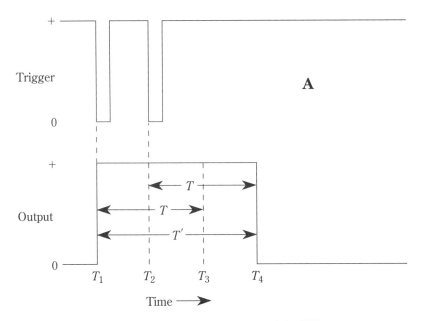

17-11A Retriggerable operation of the 555

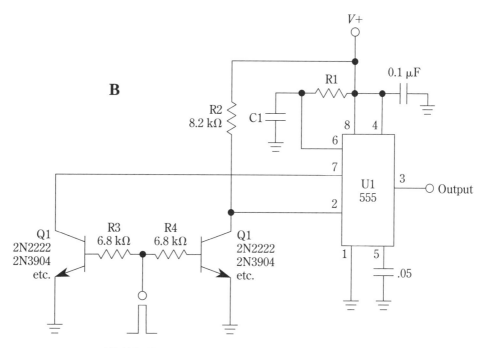

B

17-11B   Retriggerable monostable multivibrator circuit.

charge transistor seen earlier. A second transistor, Q1, is connected to the trigger input of the 555 in a manner similar to Fig. 17-7 (discussed earlier). The bases of the transistors form the trigger input. When a positive pulse is applied to the combined trigger line, both transistors become saturated. Any charge in C1 is immediately discharged, and pin 2 of the 555 is triggered by the collector of Q1 being dropped to less than $(V+)/3$. As long as no further trigger pulses are received, this circuit behaves like any other 555 MMV circuit. But if a trigger pulse is received prior to the timeout defined by the equation above, the transistors are forward-biased once again. Q1 retriggers the 555 while Q2 dumps the charge built up in the capacitor. Thus, the 555 retriggers.

Experiment 17-3

Do Experiment 17-EXP1 using a trigger circuit such as Fig. 17-11B in place of the original manual trigger circuit ($R2/S1$). Keep in mind that you will need a positive-going pulse from a square-wave generator, or a circuit such as the original trigger circuit that uses a normally closed switch for S1 (see Fig. 17-EXP2).

Astable operation of the 555 IC timer

An astable multivibrator (AMV) is a free-running circuit that produces a square-wave output signal. The 555 can be connected to produce a variable-duty-cycle AMV circuit (Fig. 17-12). A version of the circuit showing the internal stages of the 555 is shown in Fig. 17-12A, while the circuit as it normally appears in schematic drawings is shown in Fig. 17-12B. The factor that makes this circuit an AMV is that the threshold and trigger pins (6 and 2) are connected together, forcing the circuit to be self-retriggering.

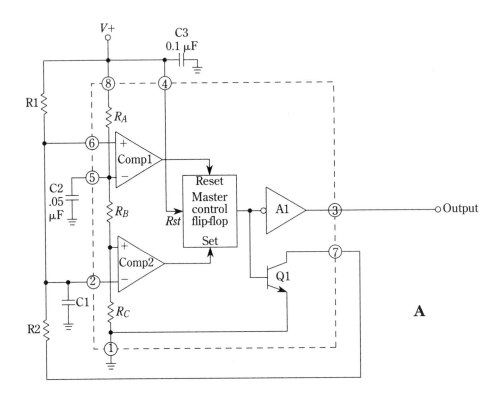

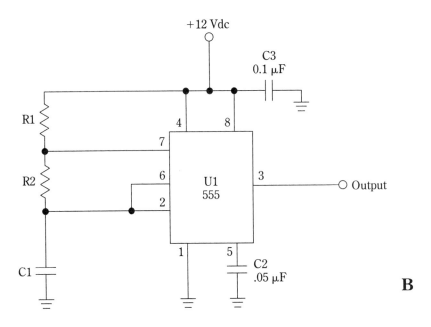

17-12 A. Astable 555 multivibrator circuit (with internal circuit shown). B. Schematic version.

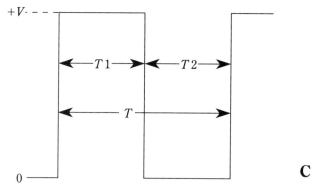

17-12C Pulse relationships.

Under initial conditions at turn-on, the voltage across timing capacitor C_1 is zero, while the biases on COMP1 and COMP2 are (as usual) set to $2(V+)/3$ and $(V+)/3$, respectively, by the internal resistor voltage divider (R_a, R_b, and R_c). The output of the 555 is high under this condition, so C_1 begins to charge through the combined resistance $[R_1 + R_2]$. On discharge, however, transistor Q1 shorts the junction of R1 and R2 to ground, so the capacitor discharges through only R2. The result is the waveform shown in Fig. 17-12C. The time that the output is high is t_1, while the LOW time is t_2. The period (T) of the output square wave is the sum of these two durations: $T = (t_1 + t_2)$. As with all similar RC-timed circuits, the equation that sets oscillating frequency is determined from the equation below:

$$T = -RC \ln\left(\frac{V - V_{C2}}{V - V_{C1}}\right) \tag{17-6}$$

For the case where the output is high (t_2 in Fig. 17-12C), the resistance R is $[R_1 + R_2]$, and the capacitance C is C_1. Because of the internal biases of the voltage-comparator stages of the 555, the capacitor will charge from $(V+)/3$ to $2(V+)/3$, and then discharge back to $(V+)/3$ on each cycle. Thus, Eq. 17.6 can be rewritten:

$$t_1 = -(R_1 + R_2)\, C_1 \ln\left(\frac{(V+) - (2(V+)/3)}{(V+) - ((V+)/3)}\right) \tag{17-7}$$

or, once the algebra is done:

$$t_1 = 0.695\,(R_1 + R_2)\, C_1 \tag{17-8}$$

By similar argument it can be shown that

$$t_1 = 0.695\, R_1 C_1 \tag{17-9}$$

For the total period T:

$$\begin{aligned} T &= t_1 + t_2 \\ &= (0.695\,(R_1 + R_2)\, C_1) + (0.695\, R_2 C_1) \end{aligned} \tag{17-10}$$

$$T = 0.695 \, (R_1 + 2R_2) \, C_1 \qquad \text{(17-11)}$$

Equation 17.11 defines the period of the output square wave. To find the frequency of oscillation, take the reciprocal of Eq. 17.11, i.e., $f = 1/T$:

$$f = \frac{1.44}{(R_1 + 2R_2) \, C_1} \qquad \text{(17-12)}$$

Experiment 17-4

This experiment demonstrates the astable configuration of the 555 IC timer device. The frequency is set low enough to permit the use of an LED output indicator circuit ($D1/R_3$ in Fig. 17-EXP4). If you wish to use an oscilloscope, then replace the output indicator with a 10-kΩ load resistor to ground.

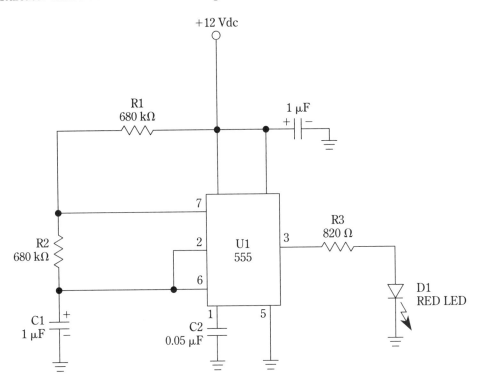

17-EXP4 Circuit for Experiment 17-4.

1. Build the circuit of Fig. 17-EXP4. Make sure that the polarity marking of C1 is observed, or else the circuit will not work properly.
2. Turn on the dc power to the circuit and observe the LED. It should be blinking on and off at a frequency determined by:

$$f = \frac{1.44}{(3 \times 680{,}000 \, \Omega) \, (1 \times 10^{-6} \, F)}$$

Make the calculation now.

3. Set the scale of a dc voltmeter to read 10 V full-scale.
4. Connect the voltmeter in parallel with capacitor C1. Use an FET-input analog voltmeter or a digital voltmeter. An ordinary analog VOM may load the circuit too much to produce a usable result.
5. Observe the capacitor charging time in conjunction with the blinking of the LED.

Duty cycle of 555 astable multivibrator

Time segments t_1 and t_2 are not equal in most cases, so the charge and discharge times for capacitor C1 are also not equal (see Fig. 17-13A). The duty cycle of the output signal is the ratio of the high period to the total period (t_1/T). Expressed as a percent:

$$\%D.C. = \frac{R_1 + R_2}{R_1 + 2R_2} \tag{17-13}$$

Various methods are used for varying the duty cycle. First, a voltage can be applied to pin 5 (control voltage). Second, a resistance can be connected from pin 5 to ground. Both of these tactics have the effect of altering the internal bias voltages applied to the comparator.

Alternatively, you can also divide the external resistances R_1 and R_2 into three values. Figure 17-13B shows a variable-duty-factor 555 AMV that uses a potentiometer (R2) to vary the ratio of the charge and discharge resistances.

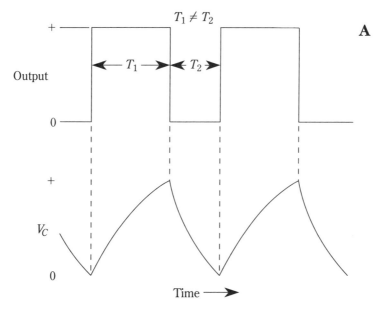

17-13A Timing relationships.

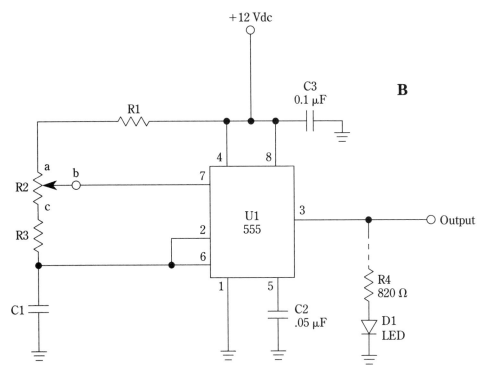

+12 Vdc

17-13B Variable duty cycle circuit.

The oscillation frequency of this circuit is found from the following equation. It is the same as the previous equations, but accounts for the resistance of the potentiometer.

$$f = \frac{1.44}{[(R_1 + R_{2ac}) + 2\,(R_{2cb} + R_3)\,C_1]}$$

Where: R_{2ab} is the resistance between terminals a and b on potentiometer $R2$; R_{2bc} is the resistance between terminals b and c on potentiometer $R2$.

Experiment 17-5

Perform Exp. 17-5 using the modified circuit shown in Fig. 17-13B. Use the following values for R1, R2, and R3:

    R1 = R3 = 470 kΩ
    R2 = 500 kΩ (multiturn is preferred)

A 555 sawtooth generator circuit

A sawtooth waveform (Fig. 17-14A) starts at a given potential (usually zero) at time t_1, and rises linearly to some value V at t_2, and then drops abruptly back to the initial

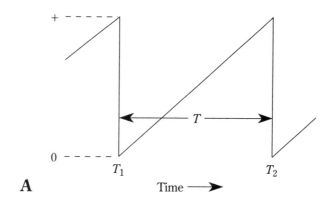

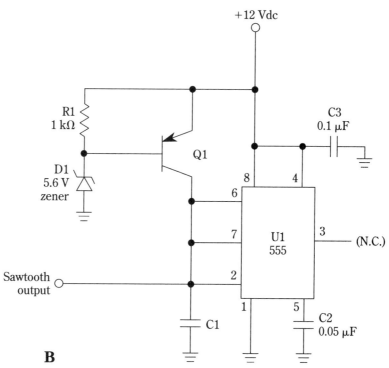

17-14 A. Sawtooth wave form. B. 555 sawtooth circuit.

condition. The circuit for a 555-based sawtooth generator (Fig. 17-14B) is simple, and is based on the 555 timer IC. The basic circuit is the monostable multivibrator configuration of the 555, in which one of the timing resistors is replaced with a transistor operated as a current source (Q1). Almost any audio small-signal pnp silicon replacement transistor can be used, although for this test the 2N3906 device was used. The zener diode is a 5.6-Vdc unit. Note that the output is taken from pins 6-7, rather than the regular chip output, pin 3, which is not used.

The circuit as shown is a one-shot multivibrator. Triggering occurs in the 555 when pin 2 is brought to a potential less than two-thirds the supply potential. When a pulse is applied to pin 2 through differentiating network R_1C_1, the device will trigger because the negative-going slope meets the triggering criteria. To make an astable sawtooth multivibrator, drive the input of this circuit with either a square wave or pulse train that produces at least one pulse for each required sawtooth. Being a nonretriggerable monostable multivibrator, the circuit of Fig. 17-14B will ignore subsequent trigger pulses during the one-shot's "refractory" period.

18
Crystal oscillator circuits

IN THE AUDIO AND LOW RF RANGE (I.E., BELOW 100 kHz), RESISTOR AND CAPACITOR (RC) elements are typically used to set the operating frequency of oscillators. But as frequency rises above 20 kHz or so into the radio-frequency (RF) range, the components of choice for frequency-setting switches to inductors and capacitors (LC) circuits. (The 20-kHz to 100-kHz region may use either RC or LC.) However, LC circuits are difficult to make with precision, and are subject to thermal drift and other problems. For operation where "rock solid" operation on a single frequency is needed, then a crystal oscillator circuit is needed. Oscillators built with crystal resonators are typically more stable than LC circuits, and can easily be built to operate on a specific frequency.

In this chapter, a project-oriented approach, rather than experiment-oriented, is used. This approach is taken mostly because these circuits are instructive in and of themselves, as well as being useful.

Piezoelectric crystals

Crystal resonators are based on the phenomenon called *piezoelectricity*, i.e., the generation of an electrical potential from mechanical deformation of the crystal surface. If a slab of the right kind of crystal at rest (Fig. 18-1A) is deformed in a certain direction, a positive potential will appear across its faces (Fig. 18-1B). Furthermore, when the same crystal is deformed in the opposite direction, the polarity of the voltage across its faces reverses (Fig. 18-1C). Thus, when the crystal is wiggled back and forth, an ac voltage appears across the faces (Fig. 18-1D).

The inverse action also occurs; when an ac voltage is applied to the faces of the crystal, it will deform in a direction determined by the polarity of the voltage (Fig. 18-2). Something special happens when the frequency produced by the oscillator matches the natural mechanical resonance of the crystal—the process becomes very

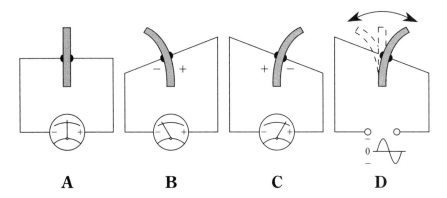

18-1 Piezoelectric crystal slab produces voltage according to how it is flexed.

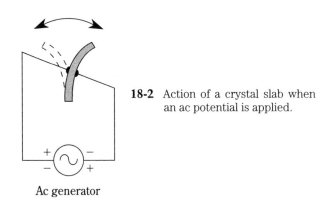

18-2 Action of a crystal slab when an ac potential is applied.

Ac generator

efficient and little energy is required to keep the motion going. This aspect of piezo-electricity is the basis for a wide range of acoustical transducers, phono pickups, and the crystal filters used in radio receiver sets, as well as oscillator frequency control.

Still another aspect to the phenomenon is also sometimes seen. When a crystal is "pinged" by a momentary pulse, it will vibrate back and forth at its resonant frequency, producing a sine-wave ac signal across its faces at that same frequency. Because of losses in the crystal, the oscillation dies out fairly quickly in an exponential decaying manner. But if the pulse that pings the crystal is repeated often enough to prevent the oscillation from dying out, then the oscillation is sustained.

These are the aspects of piezoelectricity that make it possible to use the piezo-electric crystal as a frequency-control element in an oscillator circuit. Figure 18-3A shows the usual circuit symbol for crystal resonators used in filters and oscillator circuits, while Fig. 18-3B shows two standard crystal holders.

A number of different materials exhibit piezoelectricity. Rochelle salt is a very active material that produces a large-amplitude voltage per unit of strain when mechanically deformed. However, while it is used in crystal phonograph pickups and microphones, Rochelle-salt crystal is not suitable for RF crystal oscillators. The material is very sensitive to heat, moisture, aging, and mechanical shock.

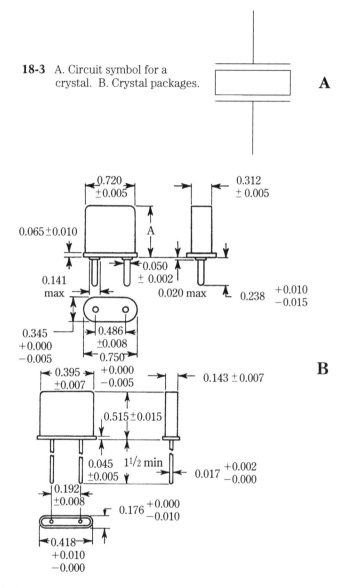

18-3 A. Circuit symbol for a
crystal. B. Crystal packages.

The next highest output material is *Tourmaline*. This material works well at all
frequencies, and it works better than other materials over the range 3 to 90 MHz.
There's only one little problem with Tourmaline: it costs an arm and a leg, as you will
discover quickly enough if you buy a Tourmaline necklace for your "significant
other." It seems that Tourmaline crystals are very popular as a variegated (red, yellow, green) semiprecious gemstone.

The best practical material for crystals is quartz. It behaves much like Tourmaline over a wide frequency range, is relatively stable, and is easily available. Although
it is used in jewelry (often mislabeled "Cape May Diamond," "Herkimer Diamond,"
and "Arkansas Diamond" in the colorless varieties, "Topaz"—which it's not; "Citrine"

is the correct name—in the yellow variety, and "Smokey Quartz" in the variety that looks like smoked glass), it is low in cost because of being plentiful. The only costly quartz crystal is the purple or violet variety, which is called Amethyst.

The quartz crystal is hexagonal in shape (Fig. 18-4), and pointed at both ends (if perfect, natural crystals are often broken or cut off on at least one end). A series of three axes are created with the Z-axis reference being from one tip point to the other; this axis is also called the *optic axis*. Crystal slabs for use as resonators are taken with different "cuts" through the crystal body. The X and Y cuts are through the X and Y axis, respectively. These are not favored for radio work, however, because they have undesirable temperature characteristics. The AT cut is made at an angle of about 35° from the Z axis. There is also a BT cut (not shown) that is sometimes used. The AT cut has a better temperature coefficient ($\Delta f/f$, ppm) by an order of magnitude, but the BT cut is usually thicker (which means it is more robust at higher frequencies where AT-cut "rocks" are very thin).

The resonant frequency of a crystal is a function of its dimensions. For example, a typical quartz-crystal resonator for 1000 kHz (1 MHz) will be approximately 0.286 cm thick and 2.54 cm square. If the crystal is ground to uniform thickness, then it will have one series resonant (f_s) and one parallel resonant (f_p) frequency. These are

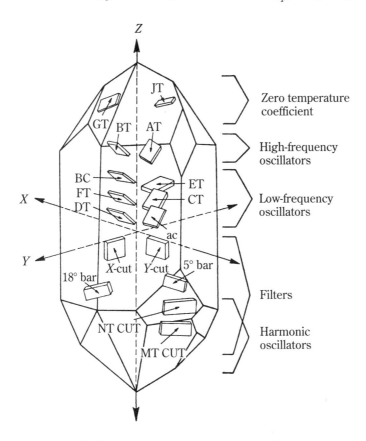

18-4 Location of various cuts of crystal.

fundamental frequencies. However, if the thickness is not uniform, then there may be spurious resonances other than the fundamental frequency.

Historically there have been two basic forms of mounting for the crystal. The older method used a pair of springs to hold a brass or silvered copper electrode against the surface of the crystal slab. World War II vintage "FT-243" crystal mounts (once popular with Novice-class hams, who were required to use crystal control on their transmitters) were of this type. Some people made "rubber crystals" by installing a pressure screw to vary the tension on the slab. These devices caused the frequency to adjust slightly, but is not a recommended practice. The other form of mount, more popular today, uses silver electrodes deposited onto the crystal surface. Wire connections can then be soldered to the surface.

The equivalent circuit for a crystal resonator is shown in Fig. 18-5A, while the reactance vs frequency characteristic is shown in Fig. 18-5B. There is a series resistance (R_s) and a series inductance (L_s) in the circuit. The series capacitance (C_s) combines with the series inductance to form a series-resonant frequency. At this frequency, because $-X_{CS}$ and $+X_{LS}$ cancel each other, the impedance of the crystal is the series resistance. That is, the impedance is minimum at the series-resonant frequency, f_s (see Fig. 18-5B). Because there is also a parallel capacitance (C_p), there will also be a parallel-resonant frequency (f_p). At this frequency, the impedance is maximum, and a 180° phase shift occurs. The parallel and series resonant frequencies are typically 1 to 15 kHz apart.

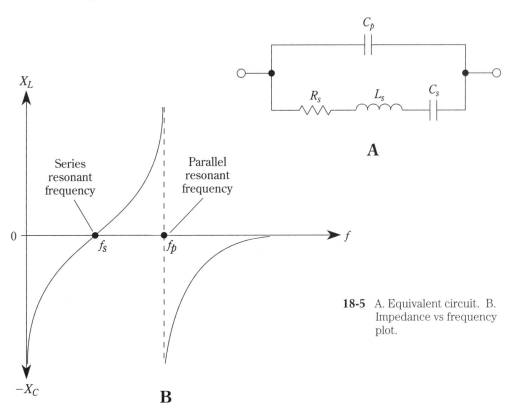

18-5 A. Equivalent circuit. B. Impedance vs frequency plot.

The design of any particular oscillator is selected to take advantage of either the series-resonant frequency or the parallel-resonant frequency. When parallel-resonant crystals are used, you must specify the load capacitance of the crystal (an external capacitance can alter the parallel-resonant frequency a small amount). Typical values are 20, 30, 50, 75, or 100 pF, although for most applications 30 (or 32) pF is specified. It is common to form a frequency adjuster by placing a small trimmer capacitor in series (Fig. 18-6A) or parallel (Fig. 18-6B) with the crystal.

When a crystal oscillator is operated at the natural series- or parallel-resonant frequency of the oscillator, it is said to be a *fundamental frequency oscillator*. The fundamental mode is used up to frequencies of 20 MHz or so. In some cases, the oscillator is operated on or near a harmonic of the fundamental frequency. This is called an *overtone oscillator*, and typically the 3rd, 5th, or 7th overtone is used over a range of 20 to 90 MHz. When ordering overtone crystals, be sure to specify the actual operating frequency, because the apparent fundamental frequency, because dividing the actual frequency of, say, a 5th overtone crystal by five does not yield the parallel-mode fundamental frequency.

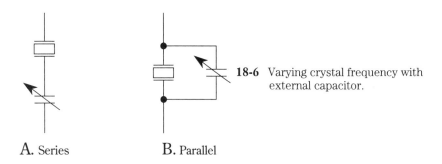

18-6 Varying crystal frequency with external capacitor.

A. Series   B. Parallel

Crystals typically want to see a certain minimum drive power in order to operate reliably, i.e., start when the circuit is turned on. But drive power can be overdone. I recall from my early ham radio days working with a fellow in Wisconsin who operated a transmitter that was a 700-W crystal oscillator. Even keeping the crystal in an oil bath wouldn't stop the fractures! It was common in those days for novice hams to build 30-W crystal-oscillator transmitters from 6L6 tubes. Crystals typically have a maximum drive power of 200 μW, although those under 1000 kHz may have maximum dissipations of 100 μW. It is common practice to operate the crystal at power levels about one-half the maximum in order to improve stability.

Fundamental mode and overtone mode

Now let's take a look at some oscillator circuits that use either fundamental-mode or overtone-mode crystals as the frequency-controlling resonator element. All of these circuits, with the exception of the TTL-based IC oscillator circuit, will work with common "garden variety" silicon transistors, JFETs, or MOSFETs.

Figure 18-7 shows a 1 to 20-MHz crystal oscillator that is easy to build and very simple. It is based on a silicon npn bipolar transistor and a crystal operated in the parallel mode. The feedback network that allows the circuit to oscillate is the capacitive voltage-divider network, consisting of C1 and C2, which is effectively shunted across the crystal (Y1). Because this circuit uses a capacitive voltage divider for the feedback network, it is a Colpitts-oscillator circuit. Capacitors C1 and C2 should be silvered mica or NPO disk ceramic types. The collector of the transistor is bypassed to ground for RF, but is at a dc potential of +5 to +15 Vdc. Output is taken from the emitter of the transistor through a 0.001-µF capacitor.

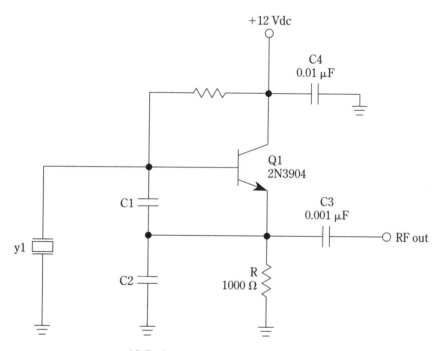

18-7 Crystal oscillator for 1–20 MHz.

The circuit shown in Fig. 18-8A is designed to operate over the range of 500 kHz to 20 MHz, depending on the values of the capacitors used in the feedback network (C_a and C_b); typical values are given in the table in Fig. 18-8A. A frequency-trimming capacitor (C_T) is provided to adjust the operating frequency to the exact required value.

The circuit will operate with superior stability and lower harmonic distortion if the feedback resistor (R1) is adjusted higher (the value can be found experimentally). However, this tactic should only be followed when the oscillator is free-running. If it is keyed or otherwise turned on and off, then a problem will be seen if R_1 is too high. Under that condition, the oscillator will not rise to its full output amplitude as rapidly as when the resistor value is lower. Dropping the resistor value lower than 2200 Ω, however, might have the effect of overdriving the crystal.

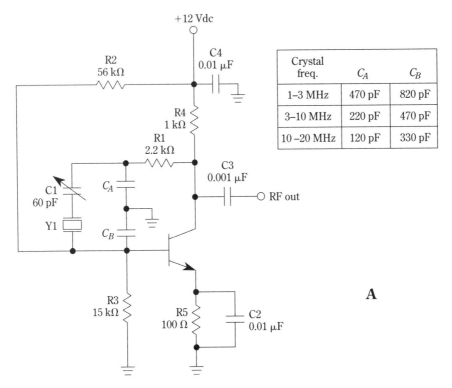

+12 Vdc

Crystal freq.	C_A	C_B
1–3 MHz	470 pF	820 pF
3–10 MHz	220 pF	470 pF
10–20 MHz	120 pF	330 pF

A

18-8A Wide range crystal oscillator.

A fundamental-frequency-oscillator circuit that is capable of providing 10 parts per million (ppm) frequency stability is shown in Fig. 18-8B. In this circuit, the crystal is connected between the emitter of the transistor and the junction point on the capacitor voltage-divider feedback network; both series-mode and parallel-mode crystals can be used. The ratio of the feedback network capacitors can be adjusted empirically for best (most stable) operation. Crystal drive level can be adjusted by making R_3 any value between 100 and 1000 Ω. The lower the value of R_3, the lower the crystal dissipation and the better the stability. Inductor L1 is resonated to the crystal frequency by C_A. The circuit will fail to start if this coil is misaligned. It is common practice to find a setting near resonance where the crystal oscillator will start reliably every time it is powered up, and has maximum stability when the circuit is operated with a buffer amplifier (discussed later).

An overtone-oscillator circuit is shown in Fig. 18-9. Although similar to the fundamental oscillator in Fig. 18-8B, it produces an output frequency at the third overtone frequency of the crystal. As with the previous circuit, crystal drive power can be adjusted by changing the value of R_1 to some value between 100 and 1000 Ω.

The variable inductor (L2) in Fig. 18-9 is resonated with C_3, and must be selected to resonate on the third overtone frequency. As in the previous case, set $L2$ to a point where the oscillator reliably starts and is stable. This coil will pull the fre-

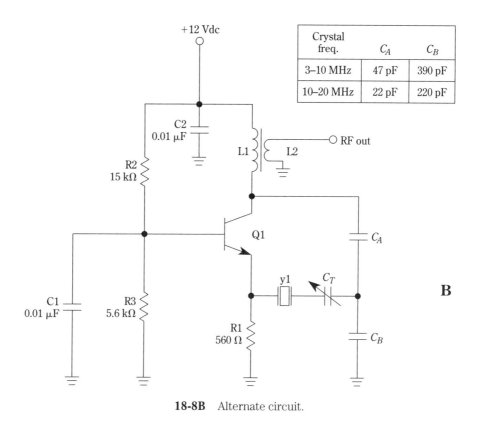

Crystal freq.	C_A	C_B
3–10 MHz	47 pF	390 pF
10–20 MHz	22 pF	220 pF

18-8B Alternate circuit.

quency somewhat, so don't adjust the frequency trimmer capacitor (C_T) the final time until after the correct adjustment point for L2 is found. After that, don't change the setting of L2. Again, a buffer amplifier is highly recommended.

A Pierce-oscillator circuit is shown in Fig. 18-10. *Pierce oscillators* are identified by having the crystal connected between the output and input of the active device. Because a junction field-effect transistor (JFET) is used in Fig. 18-10, the crystal is connected between the drain and gate; in a circuit using a bipolar transistor, the crystal is connected between the collector and base. The capacitor (C2) in series with the crystal is used in a dc blocking function. In some low-voltage transistor circuits, this capacitor can be eliminated, although most authorities would agree that it is highly desirable. Capacitor C1 is used to adjust the feedback level. For most crystals, a 100-pF unit is usually sufficient.

A pair of *Miller oscillator* circuits are shown in Fig. 18-11. The version in Fig. 18-11A uses a capacitor output circuit, while the version in Fig. 18-11B uses a link-coupled output from L1. Again, a JFET is used as the active device, even though a properly biased bipolar npn or pnp device could also be used. The Miller oscillator circuit is identified by having the crystal in a parallel-mode connection, with a parallel-resonant output-tuned tank circuit, and no capacitive voltage-divider feedback network. This circuit is quite popular, but it escapes me just why that's so. It seems

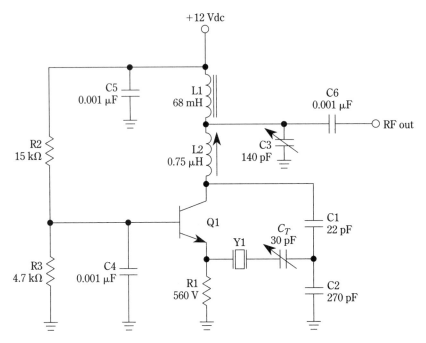

18-9 Overtone oscillator.

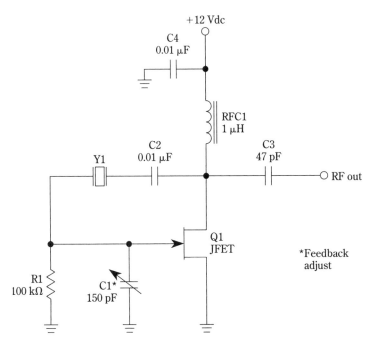

18-10 JFET Pierce oscillator.

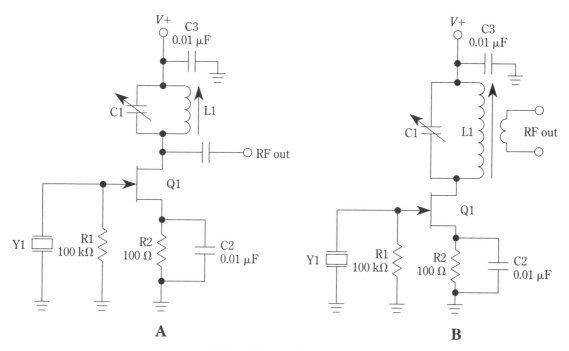

18-11 Miller-oscillator circuits.

that the Miller oscillator is subject to frequency and output-amplitude instabilities, and suffers badly from load-impedance variations. The setting of the output-tuned circuit (L_1/C_1 in both Figs. 11A and 11B) is crucial to proper operation.

A transformer-coupled-output crystal-oscillator circuit is shown in Fig. 18-12. This circuit is based on a common 2N3904 (or equivalent) npn transistor. The output transformer (T1) is wound on an FT-37-43 toroidal core. The primary consists of about 20 turns of #26 enameled wire, while the secondary is two turns of #26 wound over the primary. The primary is tapped such that one end is seven turns and the other end is 13 turns from the common point ("A"). The V+ voltage is applied to point "A," at the junction of the two halves of the primary winding.

A TTL-compatible crystal oscillator, the type used as a clock in digital circuits and computers, is shown in Fig. 18-13. This circuit can be built with any set of TTL (transistor transistor logic) inverters, although, in the case shown, the type 7400 NAND quad two-input gate is used. Because the two inputs of each gate are connected together, they are operated as inverters. Because four NAND gates are inside each 7400 package, there will be a free NAND gate for use elsewhere in the circuit if this oscillator is built. Some crystals don't like to work with this circuit, and I've experienced some difficulty making them work properly at higher frequencies (over about 10 MHz). If you want a very stable TTL clock frequency, or experience reliable starting problems, then it might be better to use one of the other fundamental-mode circuits and then convert the signal to TTL with a voltage comparator that has TTL output (i.e., an LM-311 with a 2700 pull-up resistor to +5 Vdc supply), or a 4050B or 4049B CMOS operated at +5 V, or a TTL Schmitt-trigger chip.

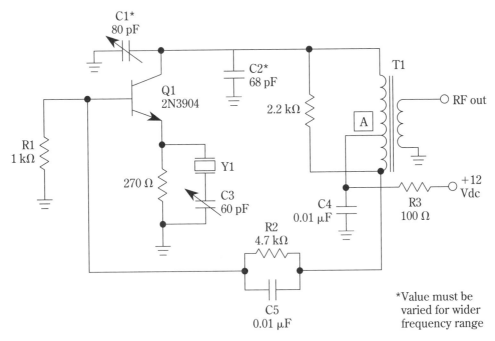

18-12 Stable HF oscillator.

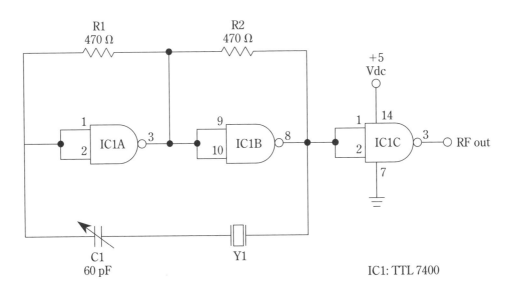

18-13 TTL crystal "clock" circuit.

Oven-controlled 1000-kHz crystal-calibrator circuit

Figure 18-14 shows the circuit for a very stable frequency standard that can be used for calibration purposes. The active devices in this circuit are a pair of metal-oxide semiconductor field-effect transistors (MOSFETs); Q1 is the oscillator while Q2 is a buffer amplifier used to isolate Q1 from the cold, cruel world. Both transistors are 3N128 MOSFETs or a service replacement type such as the NTE-220 device. These devices are static sensitive, so use "ESD" handling procedures.

The stability is enhanced in this circuit by using a separate integrated-circuit voltage regulator (IC1) to keep the voltage applied to Q1 and Q2 very stable. Variations in the voltage can cause frequency pulling and other problems, so the regulator helps substantially.

The real key to stability in this circuit, however, is the crystal oven. These devices are available from some radio parts stores, but are increasingly difficult to obtain. I've seen them sold new at hamfests, but otherwise they tend to be expensive and hard to come by—but they're worth it. The typical oven keeps the crystal at a constant temperature of 75 or 80°C. When specifying the crystal, specify a 20-pF load capacitance and inform the vendor that you need it calibrated at the oven temperature (75/80°C).

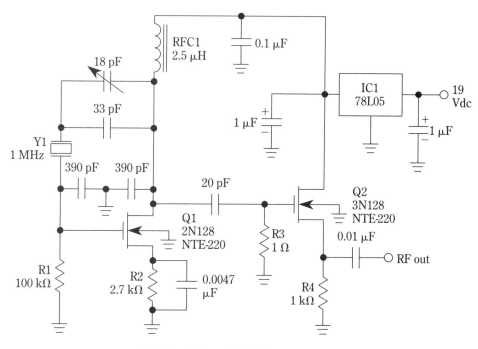

18-14 Highly stable 1000-kHz oscillator.

Accessory circuits

In this section we will take a look at a circuit that is not an oscillator, but is used with crystal oscillators. In the multichannel switch circuit of Fig. 18-15, there is more than one crystal available—but only one is used at a time. A crystal is selected by heavily forward-biasing its associated switching diode (either 1N914 or 1N4148 are used). When selector switch S1 applies +12 Vdc to a diode through a current-limiting and isolation resistor, then the affected diode is forward-biased. The crystal associated with that diode is grounded, so it starts to oscillate. Set the exact oscillation frequency by a trimmer capacitor in series with the crystal.

Buffer amplifiers are used with oscillators in order to prevent changes in the external load from pulling the oscillating frequency. Many oscillator circuits are sensitive to the load pulling effect. A *buffer amplifier* is nothing more than an amplifier stage to prevent changes in the ultimate load from being felt by the oscillator circuit. While many of the oscillators in this article will operate well without a buffer amplifier, it is always good practice to use one.

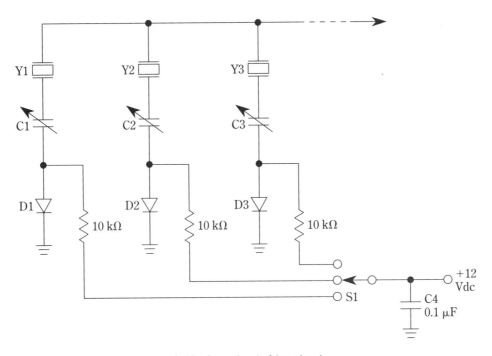

18-15 Crystal switching circuit.

Conclusion

Crystal oscillators provide a superior means of providing stable, accurate RF frequencies. They are relatively easy to build, are mostly well-behaved, and should work well.

19
Ensuring
oscillator stability

RADIO-FREQUENCY OSCILLATOR STABILITY IS IMPORTANT IN RADIO CIRCUITS, AND in both CW and (especially) single-sideband circuits it is critical. Frequency stability is one of the principal specifications that defines the quality of radio receivers and transmitters, as well as signal generators and other RF devices. A radio or signal generator that changes frequency without any help from the operator is said to drift. Frequency stability refers generally to freedom from frequency changes over a relatively short period of time (e.g., few seconds to dozens of minutes). This problem is different from *aging*, which refers to frequency change over relatively long periods of time (i.e., hertz/year) caused by aging of the components (some electronic components tend to change value with long use).

Several factors are involved in oscillator stability, and are given in this chapter as guidelines. If these guidelines are followed, it will result in a stable oscillator more often than not. For the most part, the comments below apply to both crystal oscillators and LC tuned oscillators (e.g., variable-frequency oscillators or VFOs), although in some cases one or the other is indicated by the text.

Temperature

Excessive temperatures and temperature variation has a tremendous effect on oscillator stability. Avoid locating the oscillator circuit near any source of heat within the unit. In other words, keep it away from power transistors or IC devices, rectifiers, lamps, or other sources of heat. There was one kit-built tube-type ham radio SSB transceiver (not a Heathkit, by the way) that drifted terribly because the VFO circuit was located only 2 inches or so from the vacuum tubes of the IF amplifier chain. The heat was tremendous. Using a bit of insulation to cover the VFO housing noticeably reduced the heat-induced drift of that rig!

I've seen more than a few radios come into a shop where I once worked that had ¼-inch sheets of styrofoam glued over all surfaces of the VFO housing (Fig. 19-1). In those days, the styrofoam job was nearly always a mess because only salvaged coffee cups and ¾-inch builder's styrofoam was easily available. These could be cut down to size with a hobby knife, razor knife, or other sharp tool. Today, however, art-supply stores sell a kind of poster board that makes the job real easy. This new poster board is glued to a backing of styrofoam to give it substance enough for self-support. It is very easy to cut using hobby or razor knives (I used a scalpel in some experiments). Cut the pieces to size, and then glue them to the metal surface of the shielded cabinet using contact cement or some other form of cement that will cause metal and paper to adhere together.

Operate the oscillator at as low a power level as can be tolerated so as to prevent self-heating of the active device and associated components. It is generally agreed that a power level on the order of 10 mW is sufficient. If a higher power is needed, then a buffer amplifier can be used (always a good idea, anyway).

The oscillator should be constructed in such a manner as to prevent variation in temperature. In some cases, this might mean using styrofoam or cork sheeting to line the shielded box that houses the oscillator (Fig. 19-1), while in others it might mean ensuring that the location of the box within the cabinet of the equipment served by the oscillator is at a constant temperature. At one time, it was common practice to use a constant-temperature oven for crystal oscillators. The oven kept the piezo-electric resonator at a constant temperature of 75°C or 80°C. I've only seen one, now very obsolete, piece of equipment that housed an LC VFO resonating circuit in an oven environment. In some equipment, the internal temperature will build up to a certain level and then remain stable as long as there are no air currents circulating about.

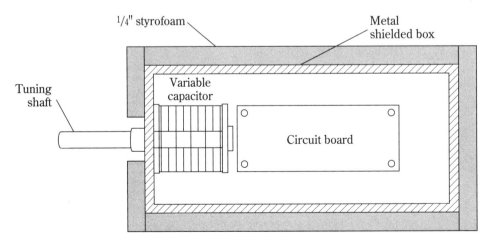

19-1 Temperature-isolated VFO chassis.

Other criteria

In general, VFOs should not be operated at frequencies above about 12 MHz. For higher frequencies, it is better to use a lower frequency VFO and heterodyne it against a crystal oscillator to produce the higher frequency. For example, one common combination uses a 5 to 5.5 MHz main VFO for all HF bands in SSB transceivers. To make a 20-meter VFO, use the 5- to 5.5-MHz VFO against a 9-MHz crystal for 14 to 14.5 MHz.

Use only as much feedback in the oscillator as is needed to ensure that the oscillator starts quickly when turned on (or keyed in the case of a CW transmitter), and does not "pull" in frequency when the load impedance changes. In some cases, a small value capacitor is inserted between the LC resonant circuit and the gate or base of the active device. The small-value capacitor prevents drift by lightly loading the tuned circuit. The most common means for doing this job is to use a 3 to 12 pF NP0 disk ceramic capacitor. However, for best stability, use an air-dielectric trimmer capacitor (2-12 pF) and adjust it for the minimum value that ensures good starting and freedom from frequency changes under varying load conditions.

A buffer amplifier, even if it is a unity-gain emitter follower, is also highly recommended. It permits building up the oscillator signal power, if that is needed, without loading the oscillator, and isolates the oscillator from variations in the output load conditions.

Power-supply voltage variations have a tendency to frequency-modulate the oscillator signal. Because dynamic circuit conditions often result in a momentary transient drop in the supply voltage, and because line-voltage variations can cause both transient drops and peaks, it is a good idea to use a voltage-regulated dc power supply on the oscillator.

It is a very good idea to use a voltage regulator to serve the oscillator alone, even if another voltage regulator is used to regulate the voltage applied to other circuits (see Fig. 19-2). Although this double-regulation approach may have been a cost burden in "long ago" times, it is now reasonable. For most low-powered oscillators, a simple low-power "L-series" (e.g., 78L06) three-terminal integrated-circuit voltage regulator is sufficient (see U2 in Fig. 19-2). The L-series devices provide up to 100 mA of current at the specified voltage, and this is enough for most oscillator circuits.

Capacitors on both input and output sides of the voltage regulator (C1 and C2 in Fig. 19-2) add further protection from noise and transients. The value of these capacitors is selected according to the amount of current drawn. The idea is to have a local supply of stored current to handle sudden demand changes temporarily, allowing time for a transient to pass, or the regulator to "catch-up" with the changed situation.

The frequency-setting components of the oscillator can also affect the stability performance. The inductor should be rigidly mounted so as to prevent vibration. While this requirement means different things to different styles of coil, it is nonetheless important.

Air-core coils are generally superior to those with either ferrite or powdered-iron cores, because the magnetic properties of the core are affected by temperature variation. Of those coils that do use cores, slug-tuned are said to be best because they can be operated with only a small amount of the tuning core actually inside the

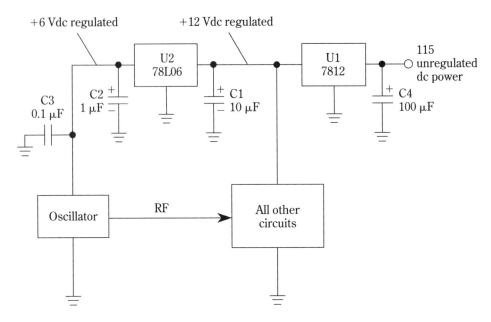

19-2 Power-supply distribution.

windings of the coil, reducing the vulnerability to temperature effects. Still, toroidal cores have a certain endearing charm, and can be used wherever the ambient temperature is relatively constant. Type SF material is said to be the best in this regard, and it is easily available as Amidon Associates. Type 6 material. For example, you could wind a T-50-6 core and expect relatively good frequency stability.

One source recommends tightly winding the coil wire onto the toroidal core, and then annealing the assembly. This means placing it in boiling water for several minutes, and then removing and allowing it to cool in ambient room air while sitting on an insulated pad. I haven't personally tried it, but the source did and reported remarkable freedom from inductor-caused thermal drift.

For most applications, especially where the temperature is relatively stable, the coil with a magnetic core can be wound from enameled wire (#20 to #32 AWG are usually specified), but for best stability it is recommended that Litz wire be used. Although a bit hard to get in small quantities, it offers superior performance over relatively wide changes in temperature. Be aware that this nickel-based wire is difficult to solder properly, so be prepared for a bit of frustration.

For air-core coils, use #22 or larger bare wire. It is probably best to use Barker & Williamson prewound air-core coils for this service. B&W makes a wide range of air-core inductors under the Airdux, Miniductor and Pi-Dux brands. Pi-Dux coils are especially suited to use in VFOs, even though intended for transmitter pi-network applications, because they have a plastic mounting plate that makes mounting easy.

I recently bought a length of the B&W Type 816A Pi-Dux product for a VFO circuit. It is 33/16 inches long, and has 16 turns per inch (16 tpi) of #18 AWG bare wire on a 1 inch diameter. The total inductance is about 17 μH, so with taps it can be used to accommodate almost any frequency within the HF spectrum. Figure 19-3 shows how the Pi-Dux coil was mounted in my project. A pair of 1-inch insulated standoffs provided adequate clearance for the coil, and held it rigidly to the chassis. The lucite mount shown in Fig. 19-3 is integral to the B&W Type 816A Pi-Dux coil. Other forms of mounting will also work, so long as it is rigid.

The trimmer capacitors used in the circuit should be air dielectric types (such as E.F. Johnson printed-circuit mounted trimmers), rather than ceramic or mica dielectric trimmers.

The small fixed capacitors used in the oscillator should be either NPO disk ceramics (i.e., zero temperature coefficient), silvered-mica or polystyrene types. Some people dislike the silvered-mica types because they tend to be a bit quirky with respect to temperature coefficient. Even out of the same batch they can have widely differing temperature coefficients on either side of zero.

Sometimes you will find fixed capacitors with other than zero temperature coefficient in a frequency-determining circuit in a temperature-compensated oscillator. The temperature coefficients of certain critical capacitors are selected to create a counterdrift that cancels out the natural drift of the circuit.

The main tuning air-variable capacitor should be an old-fashioned double-bearing (i.e., bearing surface on each end-plate) type made with either brass or iron (not aluminum) stator and rotor plates. The capacitor should be made ruggedly, and if possible use surplus military VFO capacitors. An excellent choice, where still available, is the tuning capacitors from the World War II AN/ARC-5 series of airborne transmitters and receivers.

Voltage-variable capacitance diodes (varactors) are often used today as a replacement for the main tuning capacitor. If this is done, then it becomes critical to temperature-control the environment of the oscillator. It seems that temperature variations will result in changes in diode pn junction capacitance, and that contributes much to thermal drift.

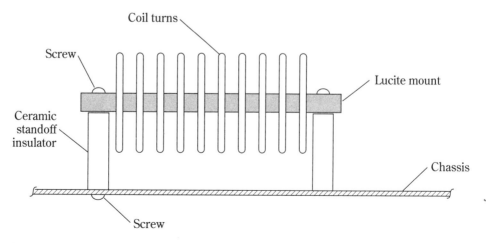

19-3 Mounting VFO coil.

Another requirement for varactor-based oscillator circuits is a clean, noise-free, separately regulated dc power supply for the tuning voltage. In most cases, you can use a low-powered three-terminal IC fixed-voltage regulator for this purpose, or alternatively an LM-317 programmed for the specific voltage.

Figure 19-4 shows a sample VFO circuit with several stability-enhancing features. This circuit is a Hartley oscillator, as identified by the fact that the feedback to the JFET transistor (Q1) is supplied by a tap on the tuning inductance (L_1). The position of this tap is usually between 10 percent and 35 percent of the total coil length. The exact position is sometimes a trade-off between stability and output power. Always opt for the stability unless it is absolutely impossible to provide a power boost (as needed) with a buffer amplifier.

The capacitors in Fig. 19-4 are selected according to the criteria discussed above. The main tuning capacitor (C1) is a 100-pF air-dielectric variable with a heavy-duty double-bearing construction. The trimmer (C2) is used to set the exact frequency, especially when a dial is used and must be calibrated. Several of the fixed capacitors are indicated with an asterisk as NP0 disk ceramic, polystyrene, or silvered mica (in order of preference).

A buffer amplifier provides two basic functions: it boosts the low output power from the oscillator to a higher level, and it isolates the oscillator from changing load impedances.

There is a voltage regulator that serves the oscillator but not the buffer amplifier. These regulators come in both metal and plastic TO-92 packages, and generate

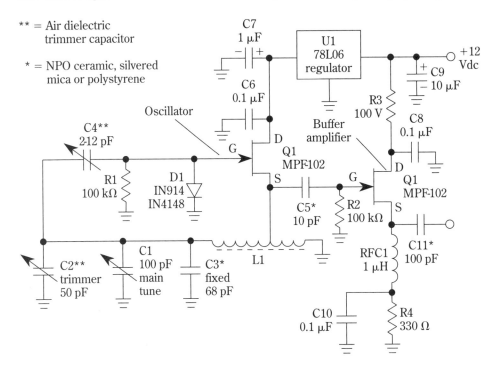

19-4 Stabilized Hartley oscillator.

very little heat. Nonetheless, it is still a good idea to mount the regulator (U1) away from the actual oscillator circuit.

The small trimmer capacitors (C2 and C4) are air-dielectric type, rather than mica or ceramic. The purpose of C4 is to provide dc blocking to the transistor gate circuit. It is such a small value because we want to load the LC tuned circuit lightly. This trimmer is adjusted from a position of minimum capacitance (i.e., with the rotor plates completely unmeshed from the stator plates), and is then advanced to a higher capacitance as the oscillator is turned on an off. The correct position is one where the oscillator starts immediately without fail every time dc power is applied.

Stabilizing varactor VFO circuits

In several chapters of this book you will find variable-frequency oscillators based on the voltage-variable capacitance diode ("varactor"). These circuits allow tuning of the oscillator frequency by applying a tuning voltage to a specially built pn junction diode. Unfortunately, varactors are temperature-sensitive, so the capacitance varies with temperature. A good way to stabilize these circuits is to supply a voltage regulator that has a temperature coefficient that is opposite the direction of the varactor drift. By matching these two, it is possible to cancel the drift of the varactor.

Figure 19-5 shows a basic circuit that accomplishes this job. Diode D1, in series with dc blocking capacitor C1, is in parallel with the inductor (L1). These components are connected into an oscillator circuit (not shown for sake of simplicity). Normally, resistor R3 (10 kΩ to 100 kΩ, typically) is used to isolate the diode from the tuning voltage source, and will be connected to the wiper of a potentiometer that varies the voltage. In this circuit, however, an MVS-460 (USA type number) or ZTK33B (European type number) varicap voltage stabilizer is present. This device is

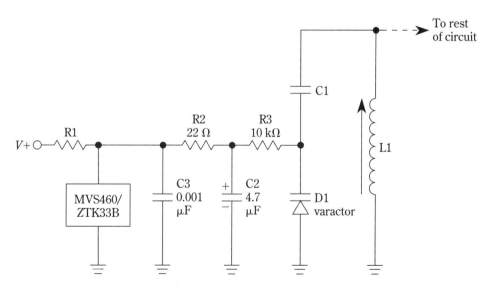

19-5 Varactor-tuned oscillator stabilization.

similar to a zener diode, but has the required –2.3 mV/°C temperature coefficient. It operates at a current of 5 mA (0.005 A) at a supply voltage of 34 to >40 V_{dc}. The value of resistor R_1 is found from:

$$R_1 \text{ (ohms)} = \frac{(V+ - 33 \text{ } V)}{0.005 \text{ } A} \qquad\qquad \textbf{(19-1)}$$

For example, when V+ is 40 V, then the resistor should be 1400Ω (use 1.5 kΩ).

Conclusion

Oscillator circuits should be stable for best operation. If you follow these guidelines, then it is likely that your circuits will be highly successful. While these techniques do not exhaust the possibilities for stable oscillator construction, they are a good start. These represent a practical collection of weapons for your use.

Appendix
Electrical safety in the electronics workshop

ELECTRICAL APPARATUS HAS BEEN AN INTEGRAL PART OF LIFE EVER SINCE MRS. Franklin told Old Ben to "go fly a kite." Devices operated by electricity make some jobs possible, and other jobs either a whole lot easier or a whole lot more accurate. But there's leaven in the electronic lump—electricity can be very dangerous. In fact, it can be fatally dangerous. Indeed, several of those who tried to duplicate Franklin's (and other early electrical) experiments were reportedly killed in the attempt. In this appendix you will learn the principal mechanisms of electrical hazards and some precautions to take, and gain some insight from some nearly tragic personal experiences of my own and others whom I know.

Electrical shock is only one of several different possible electrical hazards. In addition to electrical shock there are other electrical safety concerns. There are at least five hazard situations to consider: fire, sparking (which means possible explosions), burns, microshock, and macroshock.

Fire hazards

Fire is a major home electrical hazard, as insurance-company files will easily attest. Overloaded or defective electrical circuits can overheat and create a fire. Many fires every year are traced to faulty wiring or malfunctioning electrical equipment. One house down the block from my house has burned down twice in 20 years because of faulty electrical appliances!

Experimental circuits on the workbench are often unproven, so therefore constitute a de facto hidden fire and shock hazard. Also, things like untended soldering irons have burned down many homes and businesses over the years.

Electrical faults can also further damage the equipment that causes the problem ("secondary damage"). A short circuit that is not properly protected by either a fuse or circuit breaker may create more damage to the shorted equipment, and may also damage the building wiring and electrical components. In extreme cases, a fire may result. When fuses and circuit breakers are either not used or are defeated ("penny in the fuse box" syndrome), a severe fire hazard exists—and damage to the equipment is most certainly increased.

Sparking and explosion hazards

Explosion is often less well recognized as a hazard than simple circuit overload, but nonetheless it is highly possible under the right conditions. At least three mechanisms are possible, possibly more.

Ruptures

First, an overloaded circuit or electrical component may, depending on its design, build up internal pressure (often from gas released when the device is severely overheated) and rupture spectacularly. Examples include high-voltage transformers, batteries, and capacitors.

Sparking

The second mechanism of explosion is current-flow sparking in the presence of flammable liquids, gases, or vapors. If an electrical circuit is disconnected while operating, or if certain faults exist, then a spark may result. If that spark occurs when either flammable gases, oxygen (which isn't flammable itself, but seemingly acts that way because it promotes violent burning of other materials), or vapors (such as gasoline, certain waxes) are present, then a violent and dangerous explosion may result.

An under-the-bench "hydrogen bomb"

I once worked in a radio shop that used automobile batteries to power low-voltage dc devices being serviced on the workbench. The batteries were stored underneath the workbench on a little wooden shelf, and were charged by a trickle charger bought at an auto parts store. It worked well as a substitute for a high-current, low-voltage ac-operated dc supply. Some amateur scientists might use this method in their own laboratory. If so, please read and heed the lesson in this little story, because the battery nearly blinded a friend of mine—and could have killed him.

Automobile batteries are lead/acid storage batteries. The acid is sulphuric acid (H_2SO_4). Water has to be added to the acid from time to time, or the battery will cease working. Normal operation of these batteries produces hydrogen gas, which escapes into the atmosphere through vent holes in the water filler caps at the top of each cell of the battery. **Hydrogen gas is very explosive!**

In our shop, the batteries were serviced every Monday morning. My friend Hank (not his real name) stooped under the bench and disconnected the charger from the battery. **That was his mistake!** Immediately, a blinding flash filled the room, and our ears erupted with what sounded like a .44 Magnum "Dirty Harry" pistol going off within inches of our heads. Hank was thrown back from under the bench and, screaming, began clutching his eyes. The shop owner, my old friend the late Nelson R. Moodie, was a quick thinker and he reacted immediately. There was a garden hose in the adjacent garage (a few yards away). Nelson grabbed Hank by the shoulders and pushed him into the garage, where he turned the water hose on his eyes. Later, in the emergency room of our local hospital, Hank's clothes disintegrated from being splashed with acid. Physicians told him that only Nelson's immediate reaction saved his eyesight. His face had been within inches of the battery when the spark from the charger wire exploded the hydrogen gas collected around the battery. He escaped with a few tiny flecks of gray in the white area of his eye, and no eyesight loss!

What had Hank done wrong? First, he disconnected the charger without first turning it off, so the charger current caused a spark. If a battery is to be disconnected from the circuits it powers, then the circuits should be turned off first. Second, he failed to don the safety goggles that were provided for servicing batteries or using machine tools. Finally, if you use such a system, use a blower fan approved for explosive atmospheres to ventilate the area and disperse the hydrogen gas. Better yet, buy an ac-operated high-current dc power supply.

Hank was lucky, but don't you depend on "luck" for safety—plan ahead and avoid needing "luck!"

Besides the obvious danger of "shrapnel" from the casing of an exploding device, there is also the possibility of boiling oil or acid splattering nearby personnel. The danger from acid is obvious, but there is also a specific danger regarding the oil in some cases. Certain older capacitors and transformers were built using PCB oil as an internal coolant. PCB oil is believed to be a severe carcinogen. The importance of that statement cannot be underestimated: PCB is dangerous stuff! Although most PCB-bearing electrical devices are now largely out of service, it is possible that some are still around, especially in older equipment. Equipment to be especially suspicious of includes elderly high-power amplifiers with "oil-filled" power-supply capacitors. If one of these devices is found, then it is a good idea to have it referred to a competent expert for disposal. A PCB spill can close a building for a long period of time until a proper cleanup routine is completed.

Once, when I published a PCB warning in a magazine article, a fellow wrote to me claiming the problem is greatly overstated in the popular press. That may prove to be true, but I'll leave the issue to the environmental and medical experts—they still classify PCBs as dangerous; therefore, so will I.

Static electricity

The third hazard mechanism is like the second, except that here we deal with static electricity. The little "bite" you get grabbing the door handle of your car after sliding across the seat, or the amusing sparks seen when removing a woolen sweater in the dark, is harmless under most circumstances. But when flammable or explosive materials—especially vapors and gases—are around, then a fatal explosion can occur. Many explosions in fireworks and ammunition factories are traced to static electricity. Until about a decade ago, hospitals sometimes used flammable anesthetic gases (e.g., ether and cyclopropane) in operating rooms. Because of the very real possibility of explosion, operating rooms were designed with electrically resistive floors (5 MΩ to ground), outfitted with grounded or resistive fixtures (including carts and tables), and kept all nonexplosion-proof electrical devices more than 5 feet off the floor (the gases are heavier than air). When the flammable gases were used, the room had to be rigged for explosive gases, including conductive shoes or shoe covers for personnel, conductive liners for waste baskets, and no synthetic undergarments. Fortunately, newer nonflammable anesthetic agents have made this hazard a thing of the past in most modern hospitals.

Electrical burns

Electrical accidents can cause first-through third-degree burns. There are two basic mechanisms of burning. One (the most obvious) is the flash that is seen when an electrical arc occurs. The other occurs when current flows through tissue and causes the burning (basically, it's cooked). Ordinary 60-Hz ac household power can cause serious burns to human tissue, as any experienced emergency-room physician can testify. But radio frequency (RF) can also cause burning, and at least one physician (who is also a ham radio operator) told me that high-power RF burns tend to be more serious than 60-Hz power burns because RF burns go deeper into the tissue. I recall one radio operator who (wearing Bermuda shorts) sat on a part of his antenna wire while working on it on his basement floor. Unfortunately, he had not disconnected it from the transmitter. Someone else in the family—a small child—accidentally turned on the transmitter, sending several hundred watts of radio signal into the antenna—and my friend's legs. After the cursing and screaming was over, the fellow found second-degree burns down the back of his thigh and calf of his right leg.

Microshock

Microshock is a subtle form of electrical shock, and at one time, it was not widely recognized. Nonetheless, microshock can prove fatal under the right circumstances. Increased use of electrical equipment in hospitals in the 1950s and 1960s led some authorities to speculate that as many as 1,200 people a year were being accidentally electrocuted in hospitals from tiny electrical currents that went unnoticed by the medical staff.

Microshock is defined as a potentially dangerous electrical shock from minute currents that are too small to affect persons with intact skin, but will affect persons if they are introduced inside the body through either a wound or medical device. Microshock is not normally a problem for amateur scientists, but you should nonetheless be aware of it.[1]

In all forms of electrical shock, a difference of electrical potential (a voltage) must exist between two points on the body. In other words, two points of contact must exist between the victim and the electrical source. That is why you sometimes see harmless "hair raising" exhibits where a person touches an electrostatic high voltage (>100,000 volts!) and their hair stands on end, yet they are not shocked. These potentials are essentially "monopolar" with respect to the demonstrator, so no current flow exists. Similarly, some less prudent electricians will work a circuit "hot" (i.e., without turning off the power) in seeming safety because they take care not to ground themselves, or in any other way come between the hot wire and ground, or across two hot wires. Even so, it is an extremely unsafe practice and must be discouraged at all times! So how does this affect microshock situations?

When skin is intact, the hand-to-hand electrical resistance (i.e., opposition to the flow of electrical current) is on the order of 1000 to 20,000 Ω. However, when

---

[1]Microshock can affect animals used in science experiments. Proper experimental and animal welfare protocol calls for preventing this.

the protective skin is breached, the resistance drops to a mere 20 to 100 Ω. This phenomenon is due to the fact that internal tissue is electrically similar to salt water. Because of this lower resistance (R), a larger current (I) flows when a small voltage (V) is applied (Ohm's law states that $I = V/R$).

In addition to larger currents at smaller voltages, the lethal current threshold is lower in medical situations. By the late 1970s, medical and engineering experts set the maximum stray current level to which a patient can be subjected at 10 microamperes (μA). The question is the current density (amperes per square centimeter, A/cm^2) in the section of the heart that contains the sinoatrial (SA) node (a natural cardiac timer or "pacemaker"). Current introduced into a blood vein has a direct electrical path to the SA node, so can create a tremendous current density even though the absolute current level is small.

Combining lower resistance and lower lethal thresholds causes situations that would not normally be hazardous to become deadly in medical or veterinary situations. For example, using an ungrounded appliance or instrument (i.e., one that either lacks a three-wire power cord or has a broken ground wire in the three-wire cord) can cause leakage currents above the safe threshold level. Additional information about this type of hazard can be found in *Introduction to Biomedical Equipment Technology* by J.J. Carr and J.M. Brown (Prentice-Hall). Fortunately, starting in the early 1970s, medical-equipment designers and hospitals became more aware of electrical problems and effectively controlled the situation by improved designs.

Macroshock

Macroshock is the type of electrical shock to which all people (not just hospital patients) are subject, and comes from direct contact with an electrical source. If you touch the 110-Vac line, while grounded, then a very painful—and possibly fatal—electrical shock will occur. This event is macroshock, and does not require wounds or other breeches of the skin. Even the electrician mentioned above will get zapped badly if he slips!

How much current is fatal?

I once worked in a hospital electronics laboratory. One day I overheard an intern claim that 110 Vac from the wall socket is not dangerous because they told him in medical school that it's not the voltage that kills, it's the current. I asked him, "Doctor, have you ever heard of Ohm's law?" According to Ohm's law, the current is the quotient of voltage and resistance ($I = V/R$), so the two are related! Besides, a little statistic that the inexperienced young doctor apparently didn't know is that 110 Vac from residential wall sockets is the most common cause of electrocution in the USA. In addition, medical studies reveal that the 50- to 60-Hz frequency used in ac power distribution almost worldwide (60 Hz is used in the USA and Canada) is the most dangerous range of frequencies.

Higher and lower ac frequencies are less dangerous (but not safe!) than 60 Hz ac. Recall from above that the killing factor is current density in a certain area in the right atrium of the heart called the *sinoatrial (SA) node*. Any flow of current through the body that causes a high level of current to flow in that section of the heart can induce a fatal cardiac arrhythmia called *ventricular fibrillation (V.Fib)*, which causes death within minutes if not corrected. In general, for limb-contact electrical shocks through intact skin (macroshock), the following rules of thumb are accepted:

1 to 5 mA*	Threshold of perception
10 mA	Threshold of pain
100 mA	Severe muscular contraction
100 to 300 mA	Electrocution

(*mA = milliamperes, or $\frac{1}{1000}$ ampere)

Keep in mind that these figures are approximate, and are not to be accepted as guidelines to approximate "assumed risk." Death can occur under certain circumstances with considerably lower levels of current. For example, when you are sweating or are standing in salt water, then risks escalate tremendously. Also, certain current-flow patterns or geometries through the body are more dangerous than others, so the levels listed above may be far too low in some situations.

Mechanisms of electrical shock —some scenarios

Figure A-1 shows the schematic for the usual USA residential ac electrical system. Industrial electrical systems are a bit different at the service entrance, but become much like Fig. A-1 when the power is distributed throughout the building. The power company distributes energy through high-voltage lines. When it arrives at a point only a short distance from the customer, it is stepped down in a "pole pig" transformer (Fig. A-2) to 220 Vac center-tapped. The center tap (C.T.) of the transformer secondary is grounded, and therein lies the root of the problem. The two ends of the 220-Vac secondary are brought into the building as a pair of ground-referenced 110-Vac hot lines. Tapping across the two lines produces a 220-Vac outlet; tapping from the ground line (i.e., transformer C.T.) to either hot line produces a 110-Vac outlet.

In order to raise our consciousness about how shock can occur, let's take a look at certain scenarios of electrical shock that might occur. Figure A-3 shows the direct approach to fatal electrical shock. You are grounded through either bare feet or conductive shoes[2] and touch an electrical hot point. You need not be outdoors to be affected by this scenario. A concrete garage, shop, or basement floor is a reasonably good conductor, as are wet leather and some forms of rubber shoe.

Figure A-4 shows an indirect scenario that especially affects people using electronic instruments. Consider the grounded instrument probe (in this case an oscilloscope). When you grasp that probe, you may be grounded through the scope shield and the power-cord ground conductor. If you touch a "hot" point, then you will get shocked—and maybe killed.

A related scenario is shown in Fig. A-5. Here we have an ac/dc consumer appliance, such as a low-cost radio or TV set. Note that the oscilloscope probe ground is

---

[2]Normal shoes in good condition can become conductive when wet or if previously exposed to salty or dirty water.

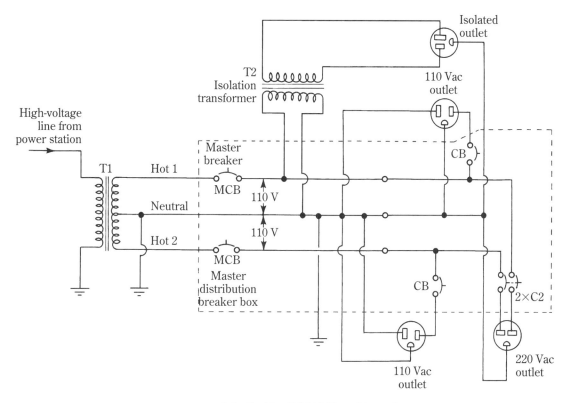

A-1 Residential 115 Vac wiring scheme.

connected to the set ground, which also happens to be one side of the ac power line. Everything is fine as long as the ac plug is oriented correctly in the wall, and if the wall socket is wired correctly. But if you plug it into the wall receptacle backwards, then there will be an explosive short circuit and possible electrocution of the operator.

Another scenario is the fatal antenna or tower erection job. It is never good practice to erect an antenna near a power line! Every year we hear stories of people electrocuted because either an antenna they were working on fell across the power lines, they tried to toss a wire antenna over the power line in order to raise the antenna above the lines, or a ladder they were using fell across the power lines. Foolish! These tactics will kill you. Incidentally, this reason is why safety-approved industrial ladders are made of wood or other nonconductive material—and not of aluminum (as are consumer ladders).

Some cures for the problems

Now that we have discussed some of the mechanisms of electrical hazards, let's take a look at some cures.

The problem that makes electrical apparatus so dangerous (especially outdoors or on concrete garage or basement floors) is that the electrical system in the USA is

A-2 "Pole pig" residential service transformer.

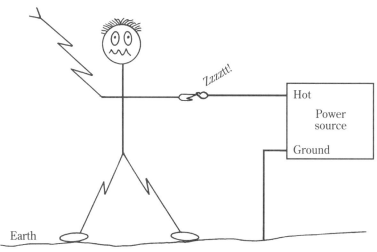

A-3 Direct contact hazard.

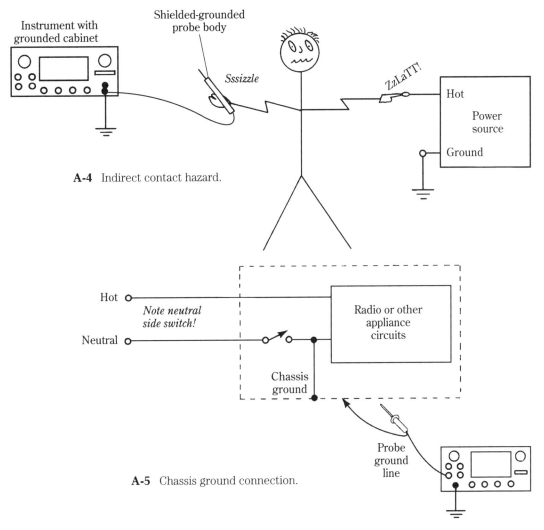

A-4 Indirect contact hazard.

A-5 Chassis ground connection.

ground-referenced (see above). The solution is to make the little "local" electrical system on your lab bench nonground-referenced. This method is used in hospital operating rooms and in some intensive-care units for patient safety reasons. It is also commonly used on radio-TV or electronic service benches, especially if ac/dc power supplies are serviced. Figure A-6 shows the wiring for such a system. Transformer T1 is one of two forms of isolation transformer: a 1:1 transformer gives a 110-Vac isolated (nonground-referenced) ac line from a 110-Vac standard line; a 2:1 transformer does the same thing from a 220-Vac line.

The second transformer in Fig. A-6, T2, is an autotransformer and is optional. It is used for varying the voltage on the ac line serving the workbench. It will typically allow you to set the output voltage from 90 Vac to 140 Vac when 110-Vac line voltage is applied.

A metal-oxide varistor ("MOV") is used to clip the amplitude of high-voltage power-line transients (lasting longer than 20 μs or so, at greater than 180 V) that

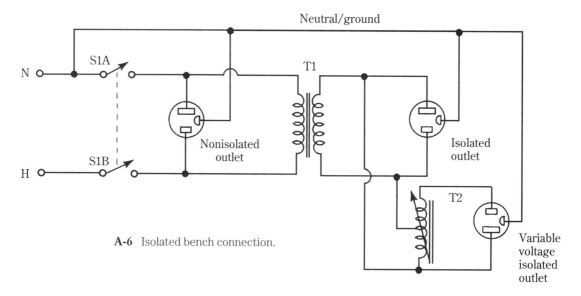

A-6 Isolated bench connection.

could either damage or interfere with the operation of the equipment on the bench. These devices are especially needed when computers or other digital electronic devices are used, and are the basis for the so-called "surge suppressor" outlet strips sold in computer stores.

A circuit breaker or fuse is used to protect equipment on the bench, as well as the transformer. It is always placed in the hot line, or in both lines. Fuses and circuit breakers are never placed in the neutral line. The switching shown in Fig. A-6 breaks both lines. I prefer this approach on the theory that hot and neutral lines can be reversed accidentally, and leave you in the position of breaking the neutral line, and leaving the hot line alive when you think it is dead!

Equipotential grounding is used in many workshops to reduce the hazard of electrical shock. An equipotential ground is a system in which all grounded points are kept at the same exact electrical potential at or near ground (0 V). In addition to being less hazardous, the equipotential ground system also works better where electronic recording devices and other instruments are used—so it has a double benefit.

To make an equipotential ground system (Fig. A-7), connect a heavy wire or copper braid between all normally grounded points, including the cabinets of instruments and appliances (Warning! Do not use ac/dc electrical devices in an equipotential ground system—in fact, don't use them at all!). This ground wire is extra to (that is, redundant to) the normal power-cord ground wire (the "third wire")

Rule to stay alive by

If you use electrical apparatus, especially outdoors or on a concrete or dirt floor, then an isolation transformer is not optional. Safety requires that it be used—and it's cheap "life insurance."

on the equipment. The method of Fig. A-7 shows star grounding (also called single-point grounding), which is superior for scientific-instrument situations. In an equipotential ground system, a short circuit to the cabinet of an instrument or other device goes to ground even if the ground wire in the cord is broken (Fig. A-8). It blows the fuse rather than blowing you out of existence.

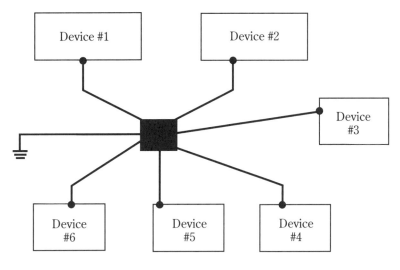

A-7 Star grounding.

Note: Power-cord grounds—the third (green) wire—often break because they are under strain. That break remains hidden because the equipment continues to operate normally. But the safety of the system is compromised when the wire is broken. That is the reason why redundant ground wires are preferred.]

A second approach to safety is just the opposite of equipotential grounding. Many modern tools or appliances that are designed for use outdoors (or on concrete floors) are double insulated. These devices have two-wire power cords, but are clearly labeled "double insulated," or in some other manner that suggests outdoor use. Don't accept the labeling, however, unless an Underwriter's Laboratory (UL) sticker is present.

Low-voltage, high-current sources—the overlooked hazard

I once attended an engineering design review meeting for a commercial marine radio transceiver. The specification called for insulation of low-voltage (<28 Vdc), high-current (>10 A) dc power-supply terminals. One of the engineers present sneered that it was something like asking him to insulate the battery terminals of his car. Implied in that contemptuous snarl is that low voltage can never hurt you. There are two false premises resident in that opinion.

First, although low-voltage, high-current points rarely cause electrical shock, it is possible for dangerous shock to occur when the person has a very low electrical skin resistance (very sweaty), or has an open wound. Although the case did not result in electrocution, one electronics technician (a friend of mine) recently injured himself severely when he cut himself on a +5-Vdc, 30-A computer power supply that

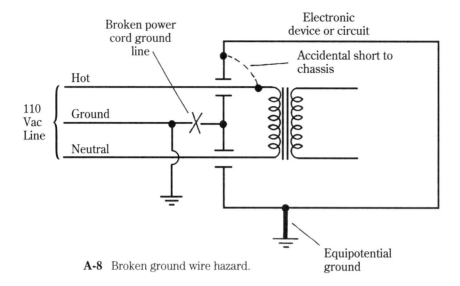

A-8 Broken ground wire hazard.

he was repairing. A large amount of current flowed in his arm, and caused severe pain and some physical damage.

Second, high current is extremely dangerous if you happen to be wearing jewelry! A two-way radio shop where I once worked used 12-V lead-acid storage batteries and battery chargers for the troubleshooting bench supply for mobile service. A technician working on the battery rack dropped a wrench, and it fell onto the battery, making contact from (−) to (+) through his watchband. The large current turned the watchband red hot, and gave him serious second- and third-degree burns on his wrist.

Don't assume that low-voltage, high-current power supplies are harmless—insulate them all!

Design your workshop for safety

An old safety cliché tells us that "safety is no accident." The truth behind the cliché is that good planning and good work habits result in a safer situation. Certain principles can be applied to your own workshop (or other venue) to avoid electrical hazards. Keep in mind that it's impossible for anyone to foresee all possible situations. Therefore, it is incumbent upon you to evaluate these recommendations for applicability, and to ensure that there are no other situations that require other methods or approaches to safety.

Master electrical cutoff switch A single electrical cutoff switch in either the hot line (or both lines if 220 Vac is used) of the ac power line should be installed. This switch should be clearly labeled "MASTER POWER CUTOFF," or something similar, and everyone in the family instructed about its use. The switch should deenergize every outlet and light on the bench except the overhead lights (how can they save your life if the room is dark). Every member of the family, including small children, should be instructed how to use this switch—and to turn off the switch immediately in the event of a mishap.

Ground fault interrupters (GFI) All ac power outlets in the work area should be of the ground-fault-interrupter (GFI) type. In most localities, the electrical code requires GFIs in wet areas (e.g., bathrooms, photo darkrooms), outdoors, or in basements and garages—so they are not optional.

Insulating floor covering The floor of the workshop should be covered with a nonconductive carpet or mat—unless explosive or highly flammable materials are handled. In dry areas, carpeting, wood, masonite, plastic, and rubber are suitable. In wet or damp areas (including any concrete floors), however, wood, carpets, masonite, and some (porous) rubber materials will absorb moisture, so are not suitable as a floor covering.

If you work with flammable or explosive materials, then consult an expert about a high-resistance grounded floor covering that will drain static electricity off harmlessly, while at the same time not putting you in a dangerous situation from ground-seeking ac power. Hospitals and some electronic assembly areas (some electronic parts are very sensitive to static voltages!) have these types of floors, so consult them on who designed their floor covering.

Isolation transformers The workbench should be equipped with an isolation transformer that has sufficient volt-ampere (e.g., wattage)[3] capacity to power all of the devices on the bench with some to spare.

Grounded ac equipment Select ac-operated instruments and apparatus that are normally grounded. Avoid ac/dc devices! The grounded variety can be identified by the three-prong (instead of two-prong) power plugs. It is also wise to redundantly ground all of the cabinets and normal ground terminals together on an equipotential ground bus.

Electrical connections Never use bare wires or uninsulated electrical connections, even in low-voltage cases. Use proper connectors—and avoid electrical tape (the cloth kind comes loose of its own accord, the plastic kind comes loose when very hot or very cold—and they both come loose in high humidity).

Ventilation If you work with potentially flammable or explosive materials, then make sure that you properly ventilate the area. Only an expert can advise on the design of ventilation for any particular material, and in most localities the fire marshal may also have an opinion or two about the matter.

Explosion-proof fixtures Working in an explosive or flammable gases or vapors environment is extremely hazardous. Normal electrical switches arc internally when activated or deactivated. Normal plugs and connectors arc momentarily when brought together. These can cause explosions! A gasoline station near my home allegedly blew up because unvented gasoline fumes were ignited by turning on the light switch on the mechanic's drop light! If you insist on working in such an environment, then use only explosion-proof switches, connectors, and appliances.

Where can an explosive environment exist in a residential structure? Well, unless you heat strictly with electricity, then you will more than likely use fuel oil, natural gas, or LP gas for energy. If those systems leak, then the structure (especially basements) will fill with an explosive mixture of gas or fumes. Also, take a look at

---

[3]Volt-amperes equals watts only in resistive circuits. If inductors, capacitors, motors, or certain other devices are in the circuit, then there is a discrepancy between VA and watts.

other activities, including hobbies, in your building. I once took an amateur jewelry-making course. One of the people using the facility told me that he used a tank of oxyacetlyene for the soldering torch needed to melt gold and silver solders. He left the master tank valve on, with a leaky handset valve, and darn near had a serious accident. It was only coincidental that he discovered the problem and called the fire department to ventilate the room with an explosion-proof fan.

Check wiring Under normal circumstances it may not be important whether or not the last electrician who wired the building or house accidentally switched the hot and neutral lines. But in amateur science labs, or anytime an appliance is used outdoors, those errors (which do happen, by the way) can be fatal! Check the wiring with an outlet tester.

What to do for an electrical-shock victim

The usual mechanism of death from electrical shock is usually a phenomenon called *ventricular fibrillation (V. Fib.)*. This is an arrhythmic heartbeat in which the heart merely quivers, instead of beating. Unfortunately, V.Fib. is not capable of sustaining blood-pumping effectiveness, so the victim dies within a few minutes unless someone trained in cardiopulmonary resuscitation (CPR) is nearby.

Before you can aid the victim of electrical shock, you must be sure that either the victim is away from the current or the current is turned off! Otherwise, when you touch the victim in order to help him or her, you will also become a victim!

As soon as the victim is clear of the electrical current, immediately initiate cardio-pulmonary resuscitation (CPR), and send for help. CPR will not bring the victim out of V.Fib. Its function is to provide life support until properly equipped and trained medical personnel can be summoned. They will use a device called a defibrillator to shock the victim's heart back into correct rhythm. They will also use drugs and intravenous (IV) solutions in order to reestablish the body's balances.

None of these actions can be performed properly by untrained people. In fact, even simple CPR cannot be effectively performed by the untrained person. Everyone who works near, on, or around electrical or electronic equipment should learn CPR. In addition, teenage and adult family members should also learn CPR; after all, who is going to save you when the electrical accident occurs at home in your laboratory, hamshack, or workshop? The local Red Cross, the Heart Association, some community colleges, and most local hospitals can direct you to certified CPR courses. It is impossible for you to learn CPR from watching "medical" or "rescue" shows on TV or in the theater, so get trained by a knowledgeable, certified instructor!

Like a hissing cobra!

Keep in mind while you read this story that I am experienced working on electrical and electronic apparatus. For more than 20 years I earned my living on the bench. Perhaps that's why I almost "bought it" one day while working on some equipment. I was working on a 110-Vac appliance in a hospital laboratory where I was employed as a technician. The equipment was live, and it was not hooked to an outlet served by the bench isolation transformer. Forgetting that it was live, I disconnected a 110-V wire and started to move it. The end of the wire hit the wrench in my other hand, and found a path to ground through my elbow. My muscles tightened up as the current coursed through my forearm. The sparking end of the wrench was "hissing like a cobra," according to a witness. I was fortunate. A friend was with me—and he yanked the cord out of the wall socket. Stupid! I knew better, but forgot the rules of good common-sense safety.

Some general safety points

There is only one way to ensure that the ac line won't shock you: disconnect it. Make it your practice never to work on equipment that has the plug inserted into the power outlet. Don't trust switches, fuses, circuit breakers, or other people. If someone were to hand you a pistol, claiming that it was unloaded, the first thing you'd do is check it yourself—the same advice also holds true for the electrical connection (which can kill you just as dead as a loaded and cocked pistol).

It is often advised that you work on electrical devices with your left hand in your pants pocket. That advice is based on the theory that the "left hand to either leg" current path is supposedly the most deadly because it passes through the heart. Even if the physiology is correct, placing one hand in your pocket leaves you awkward—and you are unable to work safely on the circuit with only one hand. It is better to use both hands, and arrange it so that the work environment is safe—and use safe technique.

What is a safe work environment? The power system should be isolated (as discussed above). The floor should be insulated by a carpet, treated masonite, a plastic cover, a rubber mat, wooden planking, or some other material; the floor should always be well-insulated and kept dry. An isolation transformer should be used on the workbench.

When working on high-voltage dc circuits, keep in mind that capacitors store electrical charge. Manually discharge all large-value (> 1 μF) capacitors after the power is turned off. Discharge the capacitor multiple times. Even when a short circuit is placed across the capacitor terminals, not all of the energy is removed the first time. Some energy is stored in the capacitor's internal dielectric material even after the main charge is discharged.

How do you short the capacitor terminals? With a screwdriver? With an alligator clip-lead? Nope! Use a shorting stick. Do not make the stick out of wood! Instead, use a dielectrically competent plastic material. Also, use more than one ground line—just in case one of them falls off while you are working, there will be one or more left to carry off the charge. Also, be sure to wear eyeglasses or safety goggles in order to prevent eye damage from flying sparks.

The only ground available to the equipment chassis when it is working on the bench is the power-cord ground. This is insufficient for safety. My workbench has a grounded 5/16-inch bolt on the back edge. When I work on electrical equipment, I ground the chassis with a heavy cable to that ground point. Don't overlook this ground.

Conclusion

There is no such a thing as complete, failure-free safety anywhere. However, proper recognition of the mechanisms of danger and proper management of the risks will ensure that your environment is as safe as possible.

Index